Technological Communities and Networks

Research has shown that technological evolution in key application areas for a competitive future increasingly requires a sustained focus on the 'invisible hands' of technological communities of actors and their variegated networks. This book seeks to look at what is distinctive about these communities, how they shape a broad range of new computer-based technologies and why they are destined to gain increased importance in explaining how these new technologies emerge.

Through analysing the structure of a broad range of existing technological communities and networks, Assimakopoulos argues that it is from collective community-based efforts rather than individual work that technological revolutions spring. Such an argument offers an implied critique of existing organisational and business models.

This book will be of great interest to research and development managers, ICT engineers and policy makers, as well as postgraduate researchers in knowledge management, technology policy, sociology, the economics of innovation and the history of science and technology.

Dimitris G. Assimakopoulos is Professor of Information Systems, Associate Dean of Research and Director of the Doctoral Program at the Grenoble Ecole de Management, France.

Routledge Advances in Management and Business Studies

1 Middle Managers in Europe
Yves Frédéric Livian and John G. Burgoyne

2 Marketing Apocalypse
Eschatology, escapology and the illusion of the end
Edited by Stephen Brown, Jim Bell and David Carson

3 Relationship Marketing in Professional Services
A study of agency–client dynamics in the advertising sector
Aino Halinen

4 Job Design and Technology
Taylorism vs anti-Taylorism
Hans D. Pruijt

5 Regulation and Organisations
International perspectives
Edited by Glenn Morgan and Lars Engwall

6 Information Technology, Organisations and People
Transformations in the UK retail financial services sector
Jeff Watkins

7 HRM, Technical Workers and the Multinational Corporation
Patrick McGovern

8 The Internationalization of Small to Medium Enterprises
The Interstratos project
Edited by Rik Donckels, Antti Haahti and Graham Hall

9 Neo-Industrial Organising
Renewal by action and knowledge formation in a project-intensive economy
Rolf A. Lundin, Hans Wirdenius, Eskil Ekstedt and Anders Soderholm

10 Perspectives on Public Relations Research
Edited by Danny Moss, Dejan Vercic and Gary Warnaby

11 Resources, Technology and Strategy
Edited by Nicolai J. Foss and Paul L. Robertson

12 Telecommunications Regulation
Culture, chaos and interdependence inside the regulatory process
Clare Hall, Colin Scott and Christopher Hood

13 Rethinking Public Relations
The spin and the substance
Kevin Moloney

14 Organisational Learning in the Automotive Sector
Penny West

15 Marketing, Morality and the Natural Environment
Andrew Crane

16 The Management of Intangibles
The organization's most valuable assets
A. Bounfour

17 Strategy Talk
A critique of the discourse of strategic management
Pete Thomas

18 Power and Influence in the Boardroom
James Kelly and John Gennard

19 Public–Private Partnerships
Theory and practice in international perspective
Stephen Osborne

20 Work and Unseen Chronic Illness
Silent voices
Margaret Vickers

21 Measuring Business Excellence
Gopal K. Kanji

22 Innovation as Strategic Reflexivity
Edited by Jon Sundbo and Lars Fuglsang

23 The Foundations of Management Knowledge
Edited by Paul Jeffcutt

24 Gender and the Public Sector
Professionals and managerial change
Edited by Jim Barry, Mike Dent and Maggie O'Neill

25 Managing Technological Development
Hakan Hakansson and Alexandra Waluszewski

26 Human Resource Management and Occupational Health and Safety
Carol Boyd

27 Business, Government and Sustainable Development
Gerard Keijzers

28 Strategic Management and Online Selling
Creating competitive advantage with intangible web goods
Susanne Royer

29 Female Entrepreneurship
Implications for education, training and policy
Edited by Nancy M. Carter, Colette Henry, Barra Ó Cinnéide and Kate Johnston

30 Managerial Competence within the Hospitality and Tourism Service Industries
Global cultural contextual analysis
John Saee

31 Innovation Diffusion in the New Economy
The tacit component
Barbara Jones and Bob Miller

32 Technological Communities and Networks
Triggers and drivers for innovation
Dimitris G. Assimakopoulos

Technological Communities and Networks

Triggers and drivers for innovation

Dimitris G. Assimakopoulos

Routledge
Taylor & Francis Group
LONDON AND NEW YORK

First published 2007
by Routledge
2 Park Square, Milton Park, Abingdon, Oxon OX14 4RN

Simultaneously published in the USA and Canada
by Routledge
270 Madison Ave, New York, NY 10016

Reprinted 2007

Routledge is an imprint of the Taylor & Francis Group, an informa business

Typeset in Times New Roman by
Book Now Ltd, London
Printed and bound in Great Britain by
MPG Books Ltd, Bodmin

British Library Cataloguing in Publication Data
A catalogue record for this book is available from the British Library

Library of Congress Cataloging in Publication Data
Assimakopoulos, Dimitris G.
Technological communities and networks: triggers and drivers for innovation / Dimitris G. Assimakopoulos.
p. cm. – (Routledge advances in management and business studies)
Includes bibliographical references and index.
1. Technological innovations. 2. Diffusion of innovations. 3. Social networks.
I. Title.

HC79.T4A817 2007
338′.064–dc22 2006027004

ISBN10: 0–415–33480–2 (hbk)
ISBN10: 0–203–41746–1 (ebk)

ISBN13: 978–0–415–33480–8 (hbk)
ISBN13: 978–0–203–41746–1 (ebk)

To Eleana, Floriane and Eleni-Anastasia:
three generations of women

Contents

Figures

Graphs

Tables

Abbreviations and acronyms

ALCS	Authors' Licensing and Collecting Society
APM	Architecture Projects Management
ARM	Advanced RISC Machines
AUT	Aristotle University of Thessaloniki
BAe	British Aerospace Defence
BP	British Petroleum
CAD	computer-aided design
CAE	computer-aided engineering
CalTech	California Institute of Technology
CAM	computer-aided manufacturing
CFD	computational fluid dynamics
CoP	community of practice
CMR	Christian Michelsen Research
DERA	Defence Evaluation Research Agency
EE/CS	electrical engineering/computer science
EC	European Commission
EU	European Union
Esprit	European Strategic Programme for Research in Information Technologies
ESRI	Environmental Systems Research Institute
FoRTH	Foundation for Research and Technology Hellas
GIS	geographic information systems
HP	Hewlett-Packard
HPC	high-performance computing
IC	integrated circuit
ICT	information and communication technologies
IST	information society technologies
MIT	Massachusetts Institute of Technology
NoP	network of practice
NTUA	National Technical University of Athens
RTD	research and technological development
SE	spectroscopic ellipsometry
SEMI	Semiconductor Equipment and Materials International

SET	secure electronic transaction
SME	small and medium-sized enterprise
STS	science and technology studies
SNA	social network analysis
SV	Silicon Valley
TAP	Telematics Applications Programme
UCB	University of California, Berkeley

Prologue and acknowledgements

This book is the outcome of the academic and professional communities and networks that I have belonged to and benefited from during the past decade or so. It owes a great debt to many people and institutions to whom I owe acknowledgements for providing inspiration and support through the three sets of research studies I carried out with the help of various collaborators in the European Union and the United States from the early 1990s onwards. A huge debt is owed to the very many people in private and public organisations, universities and government agencies who unselfishly contributed their time, wealth of ideas and insights, which enriched the overall discussion. The list is too long to reproduce here, though the ideas developed in the book mainly stem from their experiences. The responsibility for the interpretation is entirely my own and should not be attributed to any particular individual or participating/sponsoring organisation.

With respect to my doctoral research relating to the Greek GIS community, my PhD supervisor and mentor Professor Ian Masser and his team, including Professor Heather Campbell and Dr Max Craglia, in the Department of Town and Regional Planning at the University of Sheffield in the United Kingdom, deserve special thanks for providing ample guidance and enthusiasm about my research from 1992 to 1997. In this connection, I must also acknowledge the financial support provided for two years by an individual 'Marie Curie' doctoral fellowship of the European Commission's Directorate General for Science, Research and Development. Outside Sheffield, thanks are due to Professor Nikos Polydorides and his team at the University of Patras in Greece, as well as to key members of the Greek GIS community, including, Dr Thanos Doganis, Mr Adonis Kontos, Dr Poulikos Prastacos, Dr George Halaris, Dr Marinos Kavouras, Mr Dimitris Siorris and Mr Michael Salahoris, who helped to compile my initial list of contacts and facilitate this research.

With respect to my postdoctoral research relating to the informal networks of main UK contractors in Esprit, I would like to express my appreciation and gratitude to Professor Stuart Macdonald at the Sheffield University Management School (SUMS) who had the initial idea for this research and shared with patience and good humour a lot of the vagaries associated with fieldwork in the United Kingdom and the European Union. In this connection, I must also acknowledge the financial support received for two years from the UK Economic and Social Research Council through its European Context of Science Policy programme, and for one year from the EC's Esprit programme, Analysis of Socio-Economic Consequences action line, funded by DG Information Society Technologies. Dr Rebecca Piekkari from SUMS and Dr Peter Gustavsson from Linkoeping University in Sweden were also involved in the later stages of this research and greatly helped with the framing of the underlying arguments and fieldwork.

Last but not least, with respect to the research on the semiconductor community in Silicon Valley, California, I am grateful to Professor Mark Granovetter in the Department of Sociology at Stanford University, who kindly invited me to be part of his groundbreaking research group on Silicon Valley networks in 1999, and more recently in 2004. It was our collaboration with doctoral students and now colleagues, Dr Sean Everton, Dr Tsutsui Kiyoteru and Dr Emilio Castilla, that brought this research into focus and initiated the inquiry that I am pursuing in greater depth here. I would also like to acknowledge the encouragement from Professor Harry Rowen and Dr Rafiq Dossani at Stanford's Asia-Pacific Research Center, who gave me the opportunity to meet and work with Professor Martin Kenney at the University of California at Davis and Berkeley, to explore in depth the relational nature of academic entrepreneurship in the Departments of Electrical Engineering and Computer Science at Stanford and UC Berkeley.

The compilation of this book, however, is the result of the tolerance and support of colleagues in the Grenoble Ecole de Management. In particular, I am very appreciative of Dean Thierry Grange and Vice-Dean Loick Roche for allowing me the luxury of time to concentrate on and produce this monograph. The impeccable support of my series editor, Professor David Preece, and Terry Clague and his team at Routledge has also greatly helped with the production of this book.

Finally, and most importantly, I am indebted to my partner Floriane and my daughter Eleana for the love, moral support and encouragement beyond any words which they have brought to this project over the past few years. I dedicate this book to them, and to my mother Eleni-Anastasia, who inspired in me the love of learning.

1 Introduction

1.1 Communities and networks: triggers and drivers for innovation

A fundamental assumption of current strategy and policy thinking is that research and technological development (RTD) is of vital importance for innovation and competition among firms, industries and even nations (Mowery and Nelson 1999; Bresnahan and Gambardella 2004). In particular, information and communication technology (ICT) RTD is considered to be one of the great forces that has fuelled the emergence of knowledge-based economies and shaped the future of human societies in the last five decades. ICT RTD, like all computer-based technology-related work, is a social construct to a significant extent, as different actors and groups constantly shape the development of new technological ensembles (Blosch and Preece 2000) during their conception, adoption and diffusion. Diffusion research starts from the need to understand how and why a specific technological or/ and organisational innovation spreads within a social system (Rogers 2003). But sociology of technology and innovation start from the other end, trying to describe and explain how and why a specific social system, such as a technological community, or a heterogeneous network of engineers, scientists and entrepreneurs, shapes the development of an innovation and the RTD work underpinning its development (Bijker and Law 1992; Callon 1993; Latour 2005).

Ongoing research at the crossroads of the fields of technology, innovation and knowledge management; organisational theory and learning; and sociology and economics of networks suggest that technological evolution in key application areas for a competitive future, such as ICT RTD, increasingly requires a sustained focus on the 'invisible hands' of technological communities of actors and their variegated networks, which systematically produce new knowledge and attach meaning to new technologies as they originate and diffuse across heterogeneous organisational and institutional settings, disciplinary and professional groupings, as well as geographical boundaries.

This book therefore attempts to address the following topics:

- what is distinctive about technological communities and networks for illuminating the origins and different development paths of ICT technologies in heterogeneous multi-level contexts;

- how technological communities and networks shape a broad range of new computer-based ICT RTD, spanning such diverse areas of interest as semiconductors, asynchronous microprocessor architectures, Internet security, geographic information systems (GIS), electronic copyright management and intellectual property rights protection, across a large variety of organisational and institutional settings;
- why technological communities and networks are destined to gain increased importance in explaining how new technologies emerge and take different forms and functions in regional, national and international scales, against a background of the increased globalisation of basic and applied research and development activities.

In ICT RTD work, founders of start-ups, hardware and software firms, vendors, policy and decision makers from government agencies, academics and researchers from various backgrounds are some of the individual, organisational and institutional actors which form new technological communities and networks shaping the origins, evolution and diffusion of new technology. The underlying argument of this book is therefore that ICT RTD work in cross-national, national and regional scales can be analysed through the study of relevant pre-existing and new technological communities and networks, much beyond organisational boundaries and any single inventor of genius, founder or talented individual. One of the main findings is that it takes a technological community to trigger a technological revolution. The embryo of such a community is a handful of men and even fewer organisations. Technological revolutions are collective, community-based efforts, which unlike scientific revolutions are rarely the product of a single genius.

This chapter sets the scene for the book. Section 1.2 explains its research strategy by positioning this research within the fields of social informatics (Kling 2000), and science and technology studies (STS) (Bijker, Hughes and Pinch 1987; Latour 2005), also making particular reference to communities and networks of practice, technological communities and other relevant technological collaboration literature. Section 1.3 presents the book's aims and objectives, also explaining its internal organisation. Section 1.4 sets out in broad terms the research methodology for collecting, analysing and visualising network and other data for the communities and networks studied below.

1.2 A multi-theoretical and multi-level research strategy

The research strategy of the book encompasses a multi-theoretical and multi-level approach in studying technological communities and networks in a broad range of information-rich and -poor settings over the past five decades. There are three bodies of theoretical work – on technology, innovation and knowledge management; social constructivism and STS; and structural sociology and technological collaboration – that inform the discussion below (see also Chapters 2 and 3). Empirical evidence is also presented from three sets of projects at cross-national, national and regional levels of analysis (see also Chapters 4 to 6). The dual justification for pursuing such a multi-faceted research strategy is the heterogeneous process of accelerating organisational and technological change itself, related to ICT RTD work in hardware, software and services, with no clear boundaries – institutional, disciplinary or geographical – as well as the complex nature of socio-technical and techno-economic interactions underlying new ICT conception and diffusion in the European Union, the United States and worldwide.

The first body of theory that informs this research was originated in social anthropology and highlights the value of communities of practice (CoPs) for organisational learning and innovation. Etienne Wenger and Jean Lave started publishing their research (Lave and Wenger 1991) on the role of situated learning and legitimate peripheral participation in CoPs in the early 1990s. Wenger and his collaborators went on to develop his theory on CoP through the 1990s and early 2000s. Wenger (1998) is widely accepted as the seminal book on the topic. Wenger *et al.* (2002) have recently tried to capitalise on the success of the earlier work to make it useful to practising managers and business executives who are interested in managing knowledge and CoPs for fostering learning and innovation. John Seely Brown and Paul Duguid were early adopters of the CoP framework for studying organisational learning and knowledge creation in the Xerox Corporation (Brown and Duguid 1991). Their book on the social life of information (Brown and Duguid 2000a) is a significant contribution to the social informatics and knowledge management literature, since it highlights the importance of situated learning as a social process within CoPs in one of the largest technologically sophisticated and most innovative ICT companies on earth (see also Chapter 2).

The main strength of Wenger's research has to do with the innovative way in which organisations are conceptualised as constellations of informal CoPs, instead of formal functions, departments and the like. Additionally, Wenger's discussion on what are the defining characteristics of a CoP, how CoPs evolve over time, and how personal identities are formed within CoPs are extremely useful for academics and laymen alike. However, much RTD work today, especially in complex and knowledge-intensive applications areas such as ICT, is taking place across the organisational boundary (Carayannis and Alexander 1999), through inter-organisational networks (e.g. strategic alliances and joint ventures), and personal informal networks (Assimakopoulos and Macdonald 2002) connecting key people to scientific, engineering, policy and finance communities across organisational, regional and national boundaries. Wenger's CoPs seem to ignore all this RTD work, focussing mainly on the processes of working and learning 'within' the organisational boundary, and in particular, within CoPs of a simple social configuration, such as claim processors, butchers and midwives.

Moreover, CoP theory seems to be rather generic, ignoring the centrality of some communities compared to others in complex networked innovations (Swan, Scarbrough and Robertson 2002). As some commentators have observed, CoPs seem to be rather amorphous and take too lightly the power relationships among different CoPs who attach meaning to particular technologies (Fox 2000). The latter limitation of CoP theory is treated with due care within the field of STS. STS studies, and in particular social constructivism of technological systems, gained prominence in many social science disciplines, including technology and innovation management, after the publication of the volume edited by Wiebe Bijker *et al.* (1987). This seminal book put forward a number of theories and models of socio-technical and techno-economic change that shed light on the social construction of technological ensembles in a broad range of contexts. The concepts of 'relevant social groups' put forward by Pinch and Bijker (1987), and 'technological community' put forward by Edward Constant (1980, 1987) seem particularly useful for this research, since new ICTs have a programmable nature and an open architecture allowing a considerable degree of 'translation' and 'adaptation' to meet the needs of different designers, local user groups and associated technological communities (see also Chapter 2).

From the outset it seems that no matter which social groups participate in situated learning and in the subsequent shaping of a new technology this book argues that there is always an inner circle, or core of actor networks, which is associated with pre-existing and new technological communities, that critically influence the development path of a new technology, regardless of how many relevant social groups, technological communities or CoPs participate in the socio-technical processes underlying the shaping of the technology.

Edward Constant (1980), a historian of technology at Carnegie Mellon University, studied the origins of turbojet technology. He put forward the notion of a technological community, mainly based on the ideas of the historian of science Thomas Kuhn (1970). Kuhn argued that the cognitive locus of science is a well-defined community of scientists which is associated with some paradigm. Constant (1980, 1987) argues that the social locus of technological knowledge is a community of practitioners which creates and follows a technological tradition of practice associated with the evolution of a particular technology. Constant has stopped publishing on technological communities since the late 1980s. It is also worth noting that Wenger and his CoP theory do not make any reference to the earlier work by Constant. This book deliberately tries to draw the connections between CoPs and technological communities, and also takes stock of a third body of literature from structural sociology and technological collaboration, addressing head-on the first limitation of CoPs identified above.

Barry Wellman, a network sociologist based at Ontario, has been studying networked communities since the 1970s (Wellman 1979, 1999). Wellman has shifted the emphasis of the community question from inward-looking localised communities to personal communities without propinquity, based on social networks across large geographical distances. This book argues that if the notion of technological community is conceived using a social network perspective (see, for example, Wellman 1988, and Sorensen and Waguespack 2005, for a recent review of research on social networks and the organisation of research and development), since actors and linkages are the building blocks of the idea, they can be subsequently used in a meaningful way to move our thinking away from territorial definitions of community and a view of RTD based on individual geniuses associated single-handedly with particular inventions and technological innovations.

This line of research on local and global networking, and the value of cross-national and cross-regional collaboration of technical communities, has attracted considerable academic interest with respect to RTD work in new ICT in the United States and worldwide. Distinguished regional economists, such as Annalee Saxenian (2002, 2006) at Berkeley, argue convincingly for the emergence of the new 'Argonauts' in global technology regions, as far apart as Silicon Valley, Taiwan and Israel. Earlier in the 1990s, institutional economists and technology management scholars in Manchester (UK) and elsewhere (see, for example, Coombs *et al.* 1996) identified a number of the key issues for technological collaboration in RTD work and networks in industrial innovation across the firm boundary (see also Chapter 3).

The analysis and evaluation of empirical findings for this book brings together qualitative and quantitative evidence from three sets of projects at international, national and regional levels of analysis, carried out by the author and his collaborators over the past decade or so. Chapter 4 sheds light on cross-national research in ICT through the personal informal networks of Esprit (European Strategic Programme for Research in IT) main contractors in the United Kingdom and throughout the European Union in the 1990s. Chapter 5 focuses on the origins and evolution of the Greek GIS community from the early 1980s to the early 1990s, as a result of the diffusion of GIS innovations on a national scale and in its early critical stages for the shaping of GIS technology in Greece. Last but not least, Chapter 6 examines the origins and evolution of a regional technological community, the semiconductor community in Silicon Valley (SV), California, through a genealogy chart of the Semiconductor Equipment and Materials International (SEMI) from the late 1950s to the 1980s, as well as university–industry relations through academic entrepreneurship at the departments of Computer Science and Electrical Engineering at Stanford University and the University of California at Berkeley.

1.3 Aims and objectives; organisation of the book

The main assumptions underlying this book are that it is possible to gain an understanding of ICT RTD work in multiple levels of analysis, as well as the origins and evolution of new technology in its early critical stages of development, through the study of technological communities and collaboration networks across organisational, geographical, institutional and disciplinary boundaries. To this end, two main sets of objectives can be identified:

- to develop a conceptual framework in the light of the three bodies of research identified above in technology, innovation and knowledge management; organisational theory and learning; and sociology and economics of networks; for studying the notions of technological community and collaboration networks, for knowledge-intensive innovation in ICT RTD work;
- to sketch the profiles, analyse and evaluate the network structures, i.e. actors and linkages, in terms of both static and dynamics, of the technological communities related to GIS innovations in Greece, and semiconductors in SV, plus the personal networks of Esprit contractors in the United Kingdom and beyond.

In this way, whole networks of actors, people, groups and organisations who have socially constructed ICT-related innovations through RTD work in cross-national, national and regional perspectives, can be depicted, analysed and evaluated, gaining insights into the value of technological communities and networks in distributed inter-organisational ICT innovations. These network perspectives can also highlight who are the actors and groups who have instigated technological revolutions and formed the core of related technological communities and networks shaping and attaching meaning to ICT innovations in different settings and through different time periods of study over the past five decades.

Such aims and objectives have three significant advantages. They build upon existing bodies of literature, while the emphasis of the discussion is kept on the notions of technological communities and networks. They provide a heuristic way to study the processes through which a large number of actors play essentially complementary roles fostering innovation in ICT research, or/and enabling the formation of new technological traditions of practice within a whole country or region. They can also enrich current debates about the nature of ICT RTD work, shifting the emphasis of the discussion towards external sources of knowledge acquisition across organisational boundaries and social constructs, such as institutional and disciplinary settings, where new ICT is conceived, embedded and diffused. As a result, the research underlying this book shows how ICT RTD work is increasingly embedded on technological communities and networks, and shaped by a broad range of actors who come from private and public organisations, including knowledge-generating institutions such as universities.

The book's objectives are clearly located within existing socio-technical theories which assume that institutional, organisational and human issues are of significant importance with respect to effective conception, adoption and diffusion of new ICT (Davenport and Prusak 1998; Macdonald 1998; Brown and Duguid 2000a, 2001; Orlikowski and Barley 2001; Preece and Laurila 2003), and that the origins of technological revolutions can be found in the new technological communities and networks that emerge from multiple memberships (Rappa and Debackere 1992; Leonard and Sensiper 1998; Cross, Prusak and Parker 2001; Hargadon 2003; Yan and Assimakopoulos 2003) in new and pre-existing technological communities and networks related to ICT innovations. The organisation of the book into two main parts respectively addresses each of the two sets of objectives defined above. The first part – Chapters 2 and 3 – fulfils the first objective by developing a conceptual framework for the research, bringing together different strands of the theoretical perspectives related to technological communities and collaboration networks in knowledge-intensive innovation and practice.

More specifically, Chapter 2 discusses the key notions of CoPs and technological communities. It also introduces the notion of community from a network perspective, making particular reference to communities without propinquity, such as technical communities related to ICT RTD work which operate over great distances, linking engineers, scientists and entrepreneurs across organisational and geographical boundaries. This chapter also discusses the limitations of the CoP theory with respect to complex knowledge-intensive RTD work which increasingly spans organisational boundaries and involves members of a broad range of collaborating and competing scientific and professional communities and traditions of practice. One of its sections sets out the key notion of technological community by discussing the knowledge and socio-cultural dimensions of its source concept of technological tradition of practice.

Chapter 3 starts from the process of technological innovation, rather than the communities who socially shape new technologies and institutions, and it discusses in depth the increasingly important phenomenon of collaboration and networking in complex knowledge-intensive innovation. It also reviews the literature on innovation, collaboration and networking with particular emphasis on the role of formal and informal networks in complex knowledge-intensive RTD work. Its focus is kept on recent non-linear innovation models, placing the emphasis of the discussion on flexibility and speed of RTD work, 'customer focus', strategic integration with primary suppliers, and horizontal linkages such as collaborative research groupings including collaborative links with direct competitors. One of its sections discusses European innovation policies supporting collaboration in ICT RTD across organisational, institutional and national boundaries via new forms of network organisations such as collaborative projects. The discussion also highlights the importance of 'distributed capabilities', or that new and emerging technologies are less frequently produced and located within single firms, but increasingly distributed across a range of firms and other knowledge-generating institutions, especially in fast-moving areas such as ICT. Particular attention is paid here to informal networks often of a social and personal nature linking firms and other organisations, facilitating the timely and rapid flow of information for innovation.

The second part of the book presents the main findings in three chapters. Empirical research and case studies have been undertaken in a broad range of organisational settings at the cross-national (Chapter 4), national (Chapter 5) and regional (Chapter 6) scales both in the European Union and the United States. More specifically, Chapter 4 presents and analyses in a comparative framework 10 case studies on the informal networks between the main Esprit contractors in the United Kingdom and throughout the European Union. Each one of these RTD projects was supported by the European Commission (EC) within its fourth (1990–4) or fifth (1994–8) Framework programmes, bringing together a formal network of project partner organisations throughout Europe. From the outset these knowledge-intensive complex RTD projects focus on a broad range of emerging 'distributed' cutting-edge technologies related to computer hardware and software innovations. For each case study three-dimensional computer-animated maps are put together describing the socio-technical configurations of the project network and analysing both its formal and informal structures. Specialised network analysis and visualisation software (Borgatti, Everett and Freeman 2002; Richardson and Presley 2001; Batagelj and Mrvar 2004) is used to produce these maps. Finally this chapter evaluates in a comparative framework the importance of formal versus informal networks for knowledge creation and sharing in such 'distributed' and dispersed networked innovations.

Chapter 5 discusses the origins and development path of a new technological community in a national scale (i.e. Greece), adopting a longitudinal approach and focusing on a specific family of computerised technologies (i.e. GIS). The concept of a GIS community can be investigated through the existing scientific and technological communities related to the handling and analysis of geographic information, as well as the actors and their socio-technical networks which participate in the construction of GIS at local and national levels in Greece. In other words, from a sociology of technology standpoint, the question was how we could map the evolutionary path of GIS technology in Greece from its origins in the early 1980s to the development of its critical mass a decade or so later. The study of the structure of the Greek GIS community was also important for two inter-related reasons. Networks of GIS actors and linkages can highlight how information about GIS innovations spreads within a social system such as the Greek GIS community. More importantly GIS actors and linkages are the building blocks of complex socio-technical GIS networks that can drive our thinking away from territorial definitions of the GIS community concept. The study of patterns of GIS linkages within a whole country in particular can also open new research avenues for the theoretical exploration of the concept of a GIS community in different countries and cultures (Harvey 2001; Couclelis 2004).

Chapter 6 further narrows down the focus of inquiry on a regional rather than national scale. It explores the origins and emergence of the technological community of semiconductor firms in Silicon Valley, and in particular its unique informal, collaborative and entrepreneurial organisational structure and culture. Based on the SEMI genealogy chart of SV semiconductor firms, and identifying all 372 founders and 129 firms which started up in SV from 1957 to 1986, this chapter takes stock of computerised network analysis and visualisation techniques, and using this genealogy chart it demonstrates how a new horizontal network community, rather than a hierarchical workplace organisational culture, was firstly initiated with the foundation of Fairchild Semiconductor, and more importantly how this new network culture was diffused in SV through successive generations of 'Fairchildren' spin-offs up to the late-1980s. As was anticipated, the analysis of centrality and prominence of semiconductor founders and firms fuelling the origins and explosive growth of semiconductor RTD work in SV brings to light the usual suspects – Fairchild, Intel and Hewlett-Packard – but also a relatively unknown firm outside the semiconductor community, Intersil Co. It also highlights how the 'traitorous eight' founders of Fairchild instigated a new technological tradition of practice related to semiconductors and integrated circuits that gave birth to a new technological community of semiconductor firms that served as the foundation of SV's regional advantage for dominating the US and global micro-electronics industry during the past five decades.

Finally Chapter 7 summarises the main conclusions of this research with regard to the significance of technological communities and networks, and also discusses the theory-related implications with respect to the theoretical models of communities and networks of practice, technological communities and collaboration and networking. It also discusses the policy-related implications for RTD work in regional, national and cross national scales, as well as the implications of this research for future research in technological communities and networks in ICT RTD and innovation for a competitive future.

1.4 Research methodology

This research has adopted a strategy based on a social network perspective (Granovetter 1985, 2002; Rogers 1987; Wellman 1988; Wasserman and Faust 1994; Freeman, Webster and Kirke 1998; Scott 2000; de Nooy, Mrvar and Batagelj 2005) and an ethnographic approach (Hammersley and Atkinson 1983; Bryman 1988, 2004) to the study of technological communities and networks. Network scholars (see, for example, Wellman 1988) argue that it is possible to understand network phenomena, adopting either an astronomer's view of 'the universe' (i.e. the researcher is an external observer who studies 'the entire universe') or a Ptolemaic view of 'the universe' (i.e. the researcher studies particular actor networks in 'the centre of this universe'). The former approach is adopted for Chapters 5 and 6 looking into the development of relatively small technological communities at national and regional scales. The latter approach is particularly suitable for the presentation of the findings in Chapter 4, as the Esprit 'universe' is too complex to study as a whole, and the focus is kept on mapping the informal egocentric network of the main partner for each Esprit project network in a global context, comparing and contrasting it to the formal Esprit network confined among official partner organisations throughout the European Union.

The sample for the research in Chapter 4 involved all 67 Esprit projects with UK main contractors included in the Prosoma (www.prosoma.lu) showcase. Administrative leaders of these 67 projects were contacted by post and/or email and asked to identify the individual they considered to be the technological leader of their project in the UK. The findings presented in Chapter 4 are based on network data collected from 10 of these Esprit projects from 1998 to 2000. A formal project network for each main contractor was identified from the Prosoma and Cordis (www.cordis.lu) databases of the EC. Subsequently, personal informal networks were mapped following a multi-step approach. Individuals identified as technological leaders within the participating main contractors were sent postal questionnaires and each was asked to nominate up to seven other individuals (Bernard *et al.* 1984; Giusti and Georghiou 1988) who had provided information of significant value for innovation related to the specific Esprit project from within the project network itself, or any other organisation based in Europe or worldwide. In the second round, these nominated individuals were themselves contacted and asked the same question. The nomination process continued until resources were exhausted and in some cases extended to five rounds.

In addition to the multi-stage postal survey 40 semi-structured, face-to-face interviews were conducted in leading companies, universities and government agencies, including the EC itself, in the UK (Cambridge, London, Manchester, Preston), France (Grenoble, Paris, Sophia-Antipolis), Belgium (Brussels), Sweden (Stockholm, Linkoeping), Greece (Athens, Patras), and for some projects in Silicon Valley (Stanford, Los Gatos, San Francisco and Berkeley) from 1999 to 2001. It is from these that the quotations used in Chapter 4 are derived.

From the outset this research has therefore adopted a social network analysis (SNA) approach (Wasserman and Faust 1994; de Nooy, Mrvar and Batagelj 2005) in modelling networks and communities under study. SNA provides a powerful set of concepts and techniques to analyse relational data and the 'hidden' patterns of connections among actors – see also the Appendix (p. 200) for a more detailed discussion of SNA concepts and metrics used below. For example, SNA uses the notion of 'degree' to measure how many other actors directly link to a certain actor; 'centrality' to measure how critical an actor is in a network according to a measure such as 'degree'; and 'density' to measure how closely a group of actors are connected – see also the Glossary (p. 203). SNA seeks to model patterns of linkages and to describe the underlying structure of any social network or community. The power of SNA stems from its fundamental difference from non-network methodologies that ignore the underlying pattern of linkages among actors. In a non-network study, researchers often focus on the attributes of individual actors, who are viewed as isolates and under-socialised profit-maximising seekers (Granovetter 1985, 2002). From a SNA perspective the attributes of individual actors are less important than their past and ongoing relationships with other actors within the social network in which they are embedded. The behaviour of actors, say in terms of information and knowledge exchanges related to innovation, arises and is guided by structural or relational processes and norms arising out of the group and social network in which they are embedded.

For the sake of clarity, an actor or stakeholder in an informal network or/and technological community can be defined as an individual, a group or an organisation that has an interest in ICT RTD work, including GIS or semiconductor innovations, and plays a role in the conception, adoption and diffusion of relevant technological or/and organisational innovations at local, national or international scale. A linkage is a connection between two actors or stakeholders and it may take various forms:

- a formal or contractual connection (e.g. GIS software vendor and central government agency, a co-founder relation between any two founders for starting up a semiconductor firm in SV, a sub-contractor relation to the EC with respect to an Esprit project consortium);
- an informal connection between acquaintances or occasional contacts who share and exchange information about RTD work, or more generally knowledge and advice related to ICT innovations;
- and finally a personal connection of a social nature between fellow members of a private or public organisation, university, professional association, or between fellow citizens who share a common interest or are simply friends.

A network approach was adopted for this research for two additional reasons. First, it is based on the belief that regardless of whether a social system consists of a small group, a large organisation or an entire community an understanding of its social structure can facilitate (or impede) the emergence, adoption and diffusion of a technological or organisational innovation because it uncovers the 'hidden' deep structure of relations, where individuals actors are embedded. Note that network or community members, like individuals and firms, know their own linkages but they rarely know all the actors and linkages that make up a social network or community. Second, as will be discussed in Chapters 5 and 6, a key element of the socio-cultural dimension of a new technological tradition of practice is the social structure that emerges from persistent patterns of interactions, especially in the early critical stages of community development, between members of a new technological community who initiate and attach meaning to GIS and semiconductor-related innovations in different locations, and across a broad range of organisational settings.

An ethnographic and social anthropological approach was also adopted in Chapter 5 to facilitate specifically the study of the Greek GIS community from within in flexible and economical ways that allowed the researcher to evaluate his findings while doing the field research in Greece in the early 1990s. As a result of about 100 in-depth interviews and additional participant observation data from various community events during the study period, the author incrementally updated and extended his list of contacts selecting the individuals, groups and organisations who most often came up as important to meet, either because of their formal positions and relations or because of their experiences and knowledge of GIS technology. The main criteria for the choice of the people and groups forming the GIS community in Greece were their experience with GIS adoption and implementation as well as their key role in the development of GIS applications through various projects in a broad range of settings. Supporting evidence for the commitment of these individuals and groups in adopting and implementing GIS innovations in Greece, as well as their critical contribution to the development of GIS applications, was provided in a variety of contexts through their peers.

Snowball sampling of actors and linkages as suggested by Rogers and Kincaid (1981: 109) was used for the field research of the Greek GIS community which was carried out in several cities and in three different stages from 1992 to 1994. In the snowball sampling approach an original sample of respondents ('starters') are asked to name their peers, who then become the respondents in a second phase of data gathering, their contacts thus nominated becoming respondents in a third phase, etc. In this way tracing and studying the linkages is a process similar to that of a snowball rolling downhill as the sample grows slowly at the beginning and increasingly faster in later stages. The obvious advantage of the snowball sampling method is that the researcher does not arbitrarily impose the boundaries of the social system under study, but gradually uncovers them through the different responses of the participants in the research. Moreover such a sampling method provides a significant advantage to researchers who also want to use qualitative methods like participant observation, as it gradually allows both the identification and interaction with the respondents following the network linkages of the 'starters' in a multi-step sequence.

Last but not least, for Chapter 6 the SEMI genealogy chart was used to create a comprehensive database of relational data about all the co-founders and co-founded semiconductor firms in SV from the late 1950s to the late 1980s. From this secondary data set were extracted one-mode socio-matrices showing one set of actors and their linkages, e.g. the co-founder linkages among individuals or co-founded firms, as well as two-mode socio-matrices showing two sets of actors and their linkages, i.e. the main component of the

SV semiconductor community showing both the founders and founded firms (see also Chapter 6 and Appendix). In addition to this data set, based on a recent study by Kenney and Goe (2004), the author got access to Internet and postal survey data and re-classified a relational data set with respect to founder, director and advisor relationships for professorial entrepreneurship in the departments of Computer Science and Electrical Engineering at Stanford University and the University of California Berkeley (see also Assimakopoulos and Kenney 2005). Chapter 6 also discusses in more detail methods of network data analysis and visualisation wherever it is considered appropriate.

1.5 Summary

This chapter has set the scene for the book. Section 1.2 presented its multi-theoretical and multi-level research strategy. Three bodies of literature were briefly highlighted from technology, innovation and knowledge management; organisational theory and learning; and sociology and economics of networks; with pointers to the key notions of communities and networks of practice, as well as network technological communities.

The main aims and objectives, as well as the internal organisation of the book were outlined in Section 1.3. Based on the assumption that technological communities and networks can enlighten us about the origins and development paths of ICT RTD work in a broad range of information-rich and -poor settings, as well as the relationships between pre-existing and new 'revolutionary' technological traditions of practice, two sets of objectives were presented with respect to developing a conceptual framework on the one hand, and analysing and evaluating our empirical findings on the other, with regard to the personal networks of the main Esprit contractors in the UK and beyond; the origins and evolution of the Greek GIS community; and the origins and evolution of the SV semiconductor community.

Last but not least, the research methodology for this book was briefly presented in Section 1.4. From the outset, it was highlighted that the research has employed a social network approach including concepts and methods – see also the Appendix and the Glossary – that can shed light on the statics and dynamics of actors and linkages that make up the technological communities and networks under investigation.

2 Communities as the social locus of knowledge-intensive technological practice

2.1 Introduction

This section of the book discusses the conceptual framework adopted for this research. Two complementary theoretical perspectives of communities and collaboration networks in complex knowledge-intensive RTD work are presented. This chapter presents concepts of community as the social locus of knowledge in technological practice. It makes particular reference to personal network communities facilitated by communication and transportation technologies, communities of practice within organisational settings, and scientific and technological communities of practice enabling the development of existing and new technological traditions. Chapter 3 discusses collaboration and networking in RTD work starting from the processes of technological innovation, rather than the communities which socially shape new ideas and technologies. Social networks are considered important for both theoretical perspectives because they play a crucial role in understanding how knowledge about new technologies is created and shared, and because they highlight how various actors and groups give meaning to a technological innovation over great distances by reinventing and translating it during its adoption and implementation processes.

This chapter is divided into five additional sections. Section 2.2 introduces the notion of community without propinquity and also discusses personal network communities facilitated by communication and transportation technologies across large geographical distances. Section 2.3 discusses the concept of community of practice and how technology-related knowledge is often acquired through informal socialisation and situated learning in organisational and occupational communities of practitioners. Section 2.4 focuses on the relationship between scientific and technological communities also presenting the social constructivist model of technology stabilisation with particular emphasis on the concept of relevant social groups. Section 2.5 presents the concept of technological community, also discussing the notion of multiple memberships in different communities for enabling the development of new technological traditions of practice. Finally Section 2.6 is the chapter summary.

2.2 Personal network communities

Community is a concept with a long history in sociology (Tonnies 1957; Bell and Newby 1971) and other disciplines such as anthropology (Mitchell 1969) and urban planning (Park 1952). In the past sociologists defined a community primarily in terms of its spatial dimension as a territorially localised system of social relationships. In this sense, a monastery, a village or a neighbourhood is a small community of people who live together in a certain place (Sampson 1968). The critical issue which defines a community is ties between its members. For example, Weber (1947: 136) defines a communal relationship as 'a social relationship which is based on a subjective feeling of the parties, whether affective or traditional, that they belong together'.

Nisbet (1970: 47) argues that the relations between the members of a community are characterised by solidarity, intimacy, emotional depth, moral commitment, social cohesion and continuity in time. These are important issues for the well being of the individuals who belong to the community. The same issues explain why the members of the community come together, form and maintain this social system. However, it is often implicitly assumed that a social system should be enclosed, inward-looking and localised to achieve these characteristics (see for example, Nisbet 1970; Stein 1960) but this does not hold much water nowadays, and in particular is of limited usefulness for the study of new technological communities such as a new GIS community at a national scale (see Chapter 5). If the notion of community is therefore conceived using a network perspective, since actors and linkages are the building blocks of the idea they can be subsequently used in a meaningful way to move our thinking away from territorial and group-centric definitions of community (Wellman 1988).

The idea of a community without propinquity appeared in the early 1960s (Webber 1964) and shifted the emphasis of the discussion of community questions towards common values, beliefs and interests within a social network of actors, rather than spatial proximity, kinship or some kind of obligatory solidarity found in inward-looking social systems such as traditional villages. This idea of a community without propinquity is clearly illustrated in the concept of a community of scholars, or what has been called an invisible college of scientists (Crane 1972), where, for example, a professor at Stanford University in California may be better connected with other scholars in England, Japan, or Australia than his or her neighbours at Stanford, because of common ideas or scientific interests.

However, the idea of communities without propinquity is not a recent phenomenon. There were such communities in the middle ages (e.g. Benedictine monks, the masons who built the Gothic cathedrals) and even earlier in the Roman empire and the ancient Greek schools of philosophy (Popper 1998). One could argue that there have always been communities of this type. But because of increasing trends towards globalisation, communities without propinquity are now becoming much more common than previously. Perhaps in the third millennium more and more people will belong to such communities with no physical centre and clear geographical boundaries. In current jargon, these communities can be termed virtual communities (Rheingold 1994).

Network scholars such as Wellman (1979, 1999, 2002) have argued convincingly for the past few decades that community has not been lost but has been transformed into sparsely knit and loosely bound networks reflecting people's multiple memberships and identities in contemporary societies and economies. Social and communication networks connecting various actors across large geographical distances highlight this transformation (Wellman and Leighton 1979; Wellman and Wortley 1990; Wellman and Gulia 1999; Koku and Wellman 2004). People benefit from participating in such personal network communities as they can informally share scarce resources, seize opportunities across great geographical distances and learn from close friends as well as weak ties (Granovetter 1973, 1982). Transportation and communication networks and technologies including airplanes and the Internet have facilitated physical mobility and regular contacts to a significant extent, fostering socio-economic relationships locally and globally, creating what Wellman (2002) calls 'glocalised' network communities.

Saxenian's (2002, 2003, 2005) studies on global technology regions show, for example, how Taiwanese engineers, entrepreneurs and venture capitalists have successfully built and maintained personal 'glocalised' network communities connecting technical, financial and policy communities in Hsinchu, Taiwan, with their counterparts in Silicon Valley, California, strengthening the economies of both regions and reversing what not long ago was thought of as 'brain drain' to 'brain circulation'. The net result of such 'glocalised' communities is faster upgrading of the Taiwanese semiconductor and micro-electronics industries for key application areas, such as personal computers and mobile phones, fostering continuous innovation and learning across technology generations, and sustaining superior technical and business performance against intensifying global competition.

> The growing integration of the technological communities of Silicon Valley and Hsinchu offers benefits to both economies. Silicon Valley remains the center of new product definition and the developer of leading edge technologies, while Taiwan offers world class manufacturing, flexible development and integration, and access to key customers and markets in China and Southeast Asia. Unlike the arms-length and top-down technology transfers between large firms that characterized the relations between Japan and the USA in the 1980s, the Silicon Valley–Hsinchu relationship today consists of formal and informal collaborations between individual investors and entrepreneurs, small and medium-sized firms, as well as divisions of larger companies located on both sides of the Pacific.
>
> In this complex mix, the social and professional ties among Taiwanese

> engineers at home and their counterparts in the USA are often as important as more formal corporate alliances and partnerships. These relationships have been essential to establishing, maintaining, and upgrading Taiwan's role in global production networks – through OEM and ODM relationships and the myriad of other inter-firm partnerships that exploit the distinct and complementary capabilities of Silicon Valley and Hsinchu-based producers.
>
> (Saxenian 2002: 190–1)

The new 'Argonauts', as Saxenian (2006) calls these entrepreneurs and other individuals from places as far apart as Taiwan, Israel and Ireland, link small and large firms, technology markets and global production networks in an increasingly volatile and knowledge-intensive global economy. As a result of technological and economic advances coupled with political will, societies and economies open up a wide spectrum of choice for an increasing number of individuals who work in knowledge-intensive industries such as entrepreneurs and academics to establish and maintain social linkages outside their local social circles. The spatial range of support and exchange relationships has not disappeared but it has been transformed and expanded to encompass countries, continents and, in the ultimate case, the globe as a whole, fostering what Castells (1996, 2001) calls the network society and/or knowledge economy.

For example, a key objective of the EU, since 1992, has been to achieve greater economic, political, social and technological integration within and between the member states. As a result of this objective and other historical processes at the European and global scales, both individuals and organisations are able to take advantage of new opportunities to forge new multiple linkages with a significantly larger diversity of ties in a much wider framework for cooperation or competition (see, for example, the Esprit project networks in Chapter 4). Against this background of increasing globalisation it is more than ever meaningful to analyse the notion of community adopting a network approach and thinking about personal 'glocalised' network communities where the emphasis is kept not only on individual actors but also on the set of relationships that connect these actors across all sorts of geographical and institutional boundaries.

2.3 Communities of practice

Social anthropologists Lave and Wenger (1991) started publishing their research on the role of 'situated learning' through socialisation in communities of practice within organisational settings more than a decade ago. Wenger (1998) went on to develop his theory of community of practice (CoP) through the 1990s and early 2000s (see also Wenger *et al.* 2002). Lave and Wenger (1991: 98) define a CoP as 'a set of relations among persons, activity, and world, over time and in relation with other tangential and overlapping communities of practice'. The emphasis here is kept on learning and knowledge relations through mutual engagement, joint enterprise, and shared repertoire. According to Wenger (1998: 3) a CoP provides the social construct that places learning and knowledge in the 'context of our lived experience of participation in the world' as it refers to a group of people who share a common practice, have the need to share and exchange knowledge about a specific domain on an everyday basis, and are bounded by informal relationships and a shared identity within organisational boundaries.

Unlike the scholars reviewed in the previous section, Wenger and his collaborators focus on the local 'situated' nature of CoPs, within 'low-tech' apprenticeship communities. Claim processors in an insurance company, midwives, tailors, navy quartermasters and butchers were some of the communities that Wenger and his collaborators analysed in order to explain how newcomers learn from old-timers a certain work-related task embedded in the ongoing practice of a specific community. When a newcomer joins a CoP, instead of learning abstract knowledge, s/he learns to become a member of the community by developing relationships with more experienced members. First, learners participate at the periphery of community activity where they begin to assimilate and adopt the community's language, beliefs and practice. There are various paths through which learning in such occupational communities takes place. 'Changing locations and perspectives are part of an actor's learning trajectories, developing identities, and forms of membership' (Lave and Wenger 1991: 36). How much a newcomer can learn is largely determined by the level of legitimacy of his/her participation, i.e. the extent s/he is accepted by other members of the community who can provide more (or less) learning opportunities.

Above all CoP theory is therefore a social learning theory stressing the importance of social participation for generating and sharing context-specific knowledge. Moreover CoP theory sheds light on why un-codified or tacit knowledge is always 'sticky' within particular organisational contexts, since to learn more requires more social participation and acceptance by the other members of a community who can facilitate (or inhibit) situated learning and knowledge-productive relations. Wenger's research conceptualises an organisation not as a traditional set of formal functions, departments and the like, but rather as a constellation of CoPs that allow members to learn and create knowledge through mutual engagement, joint enterprise and shared experiences. This view of a firm as a learning organism consisting of multiple CoPs is a major contribution of the CoP theory and it is consistent with associated research conducted by Brown and Duguid (1991, 1998, 2000b) in one of the most innovative companies on earth, the Xerox Corporation. Since a firm is usually composed of CoPs with a certain degree of autonomy and legitimacy, these CoPs engage in new experiments, which have the potential to overcome organisational rigidities (Leonard-Barton 1992, 1995). CoPs therefore do not only foster learning but also contribute to innovation because of their constant adaptation to changing circumstances and membership. In daily work, CoPs also interact with other CoPs and are influenced by members who maintain multiple memberships in different communities (see also Section 2.5 below).

According to Wenger's analysis very often 'normal practice' does not correspond to the explicitly described functions and standard operating procedures within an organisation. Normal practice is often interpreted according to personal experiences, and the membership of one or more CoPs facilitates innovation within particular CoPs through what Brown and Duguid (1991) call 'non-canonical practice'. CoP theory therefore provides new insights into how organisations work in practice and how they learn to become more innovative by accommodating 'non-canonical practices'. In CoP theory the term 'practice' refers broadly to a shared experience (e.g. claim processing) including all the activities of a group of people who work together to accomplish a shared task, usually involving collaboration among individuals within an organisational setting or occupational community (Van Maanen and Barley 1984). Through common practice, a CoP develops a shared understanding of what it does, how members carry out everyday work, and how this particular CoP relates to and differs from other CoPs. Each CoP is therefore engaging in experiential and interpretive activities with its environment. These activities lead to a common understanding of the practice in hand, a sense of a common identity, and if managed properly they can potentially contribute to innovation with regard to various work-related practices (Wenger *et al.* 2002).

However, there are certain limitations to CoP theory which are discussed next. There has been plenty of literature over the past decade or so that has suggested that in fast-moving knowledge-intensive industries the capability of firms to access and absorb knowledge from external sources is a more important source of creating and sustaining competitive advantage than managing the existing stock of knowledge within the firms' boundaries (Badaracco 1991; Leonard-Barton 1995; Boisot 1998; Nonaka and Konno 1998; Teece 1998; DeCarolis and Deeds 1999; Fleming, King and Juda 2004). As will be discussed in detail in the next chapter much knowledge in high-technology industries flows through inter-organisational collaborative linkages such as strategic alliances and joint ventures (Osborn and Hagedoorn 1997; Hagedoorn, Link and Vonortas 2000) and also through informal personal networks (Allen 1977; Rogers 1982; Macdonald 1995, 1996; Conway and Steward 1998; Assimakopoulos and Macdonald 2003). In particular, RTD work in complex and knowledge-intensive applications areas such as ICT and biotechnology is increasingly taking place across the organisational boundary, through inter-organisational networks and personal informal networks connecting key people to scientific, engineering, policy and finance communities across organisational, disciplinary and national boundaries (see above, and also the next chapter).

Wenger's CoP theory seems to ignore all this RTD work, focusing mainly on the processes of working and learning 'within' the organisational boundary, and in particular within CoPs of a simple social configuration, such as claim processors, butchers and midwives. The focus on the relatively simple low-complexity work environments of, for example, claim processors (Wenger 1998), copier salespeople (Østerlund 1996), photocopier repairmen (Orr 1996), midwifes, butchers, tailors, and others (Lave and Wenger 1991), coupled with the methodological and epistemological demands of ethnographic cases of CoPs, may explain why the CoP appears to be a group-centric concept assuming an inward-looking, self-evolving tightly knit group, where members are closely connected but with few external linkages outside the boundaries of their organisation or workplace. Wenger (1998: 125) identifies a comprehensive list of 14 indicators for diagnosing the existence of a CoP:

1. sustained mutual relationships – harmonious or conflictual;
2. shared ways of engaging in doing things together;
3. the rapid flow of information and propagation of innovation;
4. absence of introductory preambles, as if conversations and interactions were merely the continuation of an ongoing process;
5. very quick set-up of a problem to be discussed;
6. substantial overlap in participants' descriptions of who belongs;
7. knowing what others know, what they can do, and how they can contribute to an enterprise;
8. mutually defining identities;
9. the ability to assess the appropriateness of actions and products;
10. specific tools, representations and other artefacts;
11. local lore, shared stories, insider jokes, knowing laughter;
12. jargon and shortcuts to communication as well as the ease of producing new ones;
13. certain styles recognised as displaying membership;
14. a shared discourse reflecting a certain perspective on the world.

This set of indicators can only be fulfilled by a limited number of CoPs, most often situated in a common location such as an apprentice-based workplace or office. Many large and small firms, however, organise their RTD work in distributed workplaces (e.g. Lam 1997, 2003). Colleagues who have close working relations are not always located in a single place or organisation. Work can take place by phone calls, emails, the electronic exchange of documents, and face-to-face interactions in diverse settings. Recently researchers have therefore tried to expand the CoP theory into inter-organisational settings: for example, in the work of distributed communities of practice across international boundaries (Hildreth and Kimble 2000) and virtual communities of practice (Kimble, Hildreth and Wright 2001).

Brown and Duguid (2001: 198) have also recently pointed out that 'often too much attention is paid to the idea of community, too little to the implications of practice'. Their analysis of practice has highlighted the epistemic differences across different practices and the significance of CoP theory in explaining how knowledge is not only embedded in CoPs within organisational boundaries and occupational communities (Van Maanen and Barley 1984), but also networks of practice (NoPs) that extend beyond the organisational boundary (Brown and Duguid 2001). Because of shared practice Brown and Duguid (2001) argue that knowledge can be both 'sticky' and 'leaky'. Individuals who share a common practice can easily communicate within organisational boundaries with other members of the local CoP. Moreover, knowledge related to a common practice can easily be shared within a broader network of practitioners, as well as collective groupings, such as professional associations (Swan *et al.* 1999) or specialised online communities of interest (Assimakopoulos and Yan 2006b).

Last but not least, CoP theory seems to be rather generic, ignoring the centrality of some CoPs compared to others in complex networked innovations. See, for example, Swan *et al.* (2002) who highlight the centrality of one group of medical doctors (i.e. surgeons) over other relevant professional groups (e.g. urologists) in the process of shaping a new treatment of prostate cancer. As Fox (2000) and other commentators (e.g. Gherardi 2001) have observed, CoP theory seems to be rather amorphous and takes too lightly the power relationships among different CoPs who attach meaning to particular technologies. The latter limitation of CoP theory is treated with due care within the field of science and technology studies (STS) as is discussed next.

2.4 Communities in science and the science/technology relationship

From a historical perspective, the social organisation of scientific progress assumes a community structure, as Kuhn (especially 1970: 174–210) and others (Price 1963; Merton 1973; Pickering 1995; Knorr-Cetina 1999) have asserted. Crane (1972: 63) argues that since the seventeenth century scientists have pioneered the development of communication networks which were not confined to local social circles. They have transcended national boundaries, promoting the exchange and sharing of ideas as well as cross-fertilising different scientific fields. See, for a recent example, Verspagen and Werker (2003) who analyse the invisible college of the economics of innovation and technological change. Crane (1972: 22) also argues that the growth of scientific knowledge is a kind of diffusion process in which ideas are communicated from person to person within or between scientific communities of interest. As a result, the cumulative number of publications for any scientific field follows the logistic growth curve used to depict the diffusion of innovations (see Figure 2.1, put forward by Rogers 2003).

As can be seen from Figure 2.1, in stage 1 when a paradigm – 'a successful piece of science that serves a community of scientists as a model for future work' (Gutting 1984: 49) – appears there are very few publications and little

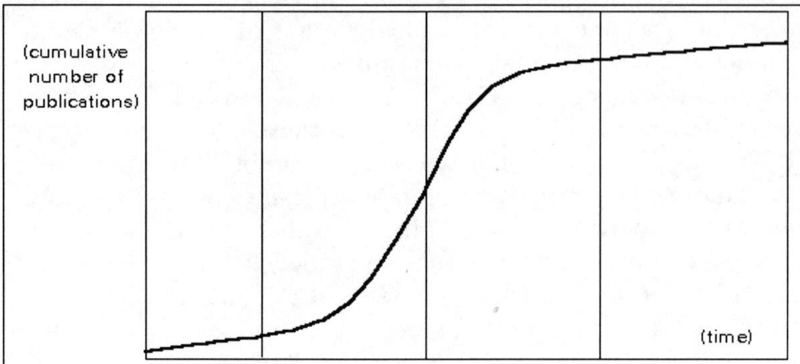

	Stage 1	Stage 2	Stage 3	Stage 4
CHARACTE-RISTICS OF KNOWLEDGE	Paradigm appears	Normal science	Solution of major problems	Exhaustion
			Anomalies appear	Crisis
CHARACTE-RISTICS OF SCIENTIFIC COMMUNI-TIES	Little or no social organi-sation	Groups of collaborators and an invisible college	Increasing specialisation	Decline in membership
			Increasing controversy	Decline in membership

Figure 2.1 Characteristics of scientific knowledge and of scientific communities at different stages of the S-shaped logistic curve.

Source: Crane 1972: 172.

social contact among scientists who work within this particular area of interest. For example, when Einstein wrote his theory of general relativity there were very few publications about this kind of problem and very few scientists shared his interest and ideas (Schwartz and McGuinness 1992). In stage 2, the number of publications increases rapidly, as small groups of collaborators set up an invisible college, or scientific community, to exchange ideas, share experiences and subsequently communicate the results. Normal scientific practice occurs at this stage of development of the scientific community as its members develop new theories according to the new paradigm and publish in academic journals and books. In stage 3 the number of publications is still growing but most of the major problems associated with this paradigm have been tackled and the pace of growth in the cumulative number of publications is decreasing. Explanatory difficulties or the failure of scientific theories (i.e. anomalies) developed on the basis of the paradigm usually generate controversy among members of the invisible college. Increased specialisation causes a decline in membership, and a paradigm which is overwhelmed by anomalies is usually replaced with a new paradigm.

A handful of scientists of the invisible college set up in stage 2, together with a number of outsiders, usually initiate a new scientific revolution and set up a new invisible college or scientific community to undertake normal science according to a new paradigm. In stage 4 the paradigm that appeared in stage 2 is exhausted. There is a crisis and a further decline in membership of the invisible college formed in stage 2 (see Figure 2.1).

Until the early 1970s most of the research interest was limited to science and scientific communities (Kuhn 1970; Crane 1972). The main assumption underlying the literature was that science generates the knowledge employed by technologists and that the relationship between science and technology is unidirectional (Layton 1972, 1977) resulting in a linear model of innovation, where scientists carry out 'basic' research and produce knowledge that is subsequently used by technologists for RTD work and downstream commercialisation. At the time, three decades ago, the field of history of technology, in contrast to the field of history of science, was very new (Hughes 1979). Pinch and Bijker (1987: 17) argue that the separation of science from technology continued until the late 1980s, although the need for a unified approach to science and technology was evident earlier. For example, Layton (1977: 210) argued that:

> Science and technology have become intermixed. Modern technology involves scientists who 'do' technology and technologists who function as scientists ... The old view that basic science generates all the knowledge which technologists then apply will simply not help in understanding contemporary technology.

Nowadays scientists and technologists are increasingly enmeshed in a symbiotic relationship as academics increase collaboration with practitioners who share the same kind of problems and compete for the same resources. As will be discussed in detail in the next chapter, a number of non-linear innovation models in the past decade or so have highlighted this shift in the macro-innovation environment: e.g. Gibbons *et al.* (1994) in terms of the dynamics of science and the distributed nature of new knowledge production and Etzkowitz and Leydesdorff (2000) in terms of a triple helix in university, industry government relationships.

The boundaries between science and technology are increasingly blurred today as university-based scientists, together with various practitioners and other actors from private and public organisations, increasingly create and share knowledge participating in the construction of new technologies (see, for example, the Esprit and IST programmes of the European Commission which have as a main objective to support and enhance the interactions among large and small firms, universities and government research institutes in relation to RDT work for ICT). As a result, heterogeneous networks among a broad range of actors are substantially encouraged since the bringing of different people and institutions together can provide more flexible and effective organisational structures for the sharing of ideas and the solving of scientific and technological problems.

For these reasons, Geisler (1993: 335) argued more than a decade ago that the field of university–industry–government relations had witnessed a quiet revolution in thought and in action since the early 1980s. Global cooperation between universities, industry and government has gained increased prominence (see also Assimakopoulos 2003). Academics, private sector managers and government officials are not only promoting such cooperation, but also rethinking their attitudes and approach to this phenomenon. Science and technology have a two-way relationship. Science informs and shapes technology but technology also sets the agenda and provides the stimulus for much scientific research. As a result, the deterministic view of the impact that science or technology has on society has been seriously challenged, with sociologists of technology (e.g. Callon and Latour 1981; Callon and Law 1982; Latour 1991, 2005; Law 1994) arguing that even the technical meaning of hard technology is socially constructed.

Under these circumstances almost everything is negotiable: who is a scientist and who is a technologist, what is technological and what is social, what is certain and what is not, who in the ultimate case can participate in the controversy of defining what a new technology can and cannot do. In this sense technology, like science, is a social construct. Therefore an interesting question is to find models and concepts which can explain the social construction of technological systems. New technological systems are open to different interpretations from different actors and social groups. This is particularly true for new computer-based technologies which have a programmable nature and an open architecture allowing a considerable degree of reinvention and adaptation to meet the needs of different designers and user groups: see, for example, Rogers (1993) and Campbell and Masser (1995) with regard to GIS innovations discussed in Chapter 5. In this respect two different concepts of socio-technical change are of particular importance. These are the social constructivistic model of technological stabilisation and the concept of the technological community. The former is discussed below and the latter is discussed in the next section.

Pinch and Bijker's (1987) social constructivistic model of the evolution of a new technology makes use of the main concept of relevant social group to explain how artefacts take the form that they do. They argue that different social groups have different interests in the development of a technical artefact. As a result, in the early stages of the evolution of an artefact, alternative designs are produced which try to solve different problems and fulfil different needs. In later stages of technological development, because of social, technical, economic and political constraints, there is an increasing degree of stabilisation as many of these designs disappear in the competition for survival, while a few of them survive and become successful. The concept of the relevant social group in the evolution of a technical artefact denotes institutions and organisations as well as organised or unorganised groups of individuals who share the same set of meanings in relation to an artefact (Pinch and Bijker 1987: 30). The role of relevant social groups in the shaping of a new technology is critical, since the problems and solutions which are identified and addressed by each group highlight how various actors participate in the development process of a new technology.

The application of the social constructivistic model of technological stabilisation is best illustrated by reference to the evolution of the air tyre in bicycle technology which has been studied from this standpoint by Pinch and Bijker (1987) (see also Bijker 1995; Rosen 2002). For bicycle technology the relevant social groups were not only its producers and users but also those who opposed the adoption of specific features on bicycles such as the air tyre on technical or aesthetic grounds. Among the users may be distinguished men and women, older and younger cyclists, as well as sports cyclists and tourist cyclists. The air tyre, which is considered a standard feature of the bicycle today, created a lot of controversy back in the 1880s and 1890s. In the beginning it was considered a solution to the vibration problem in other categories of vehicles such as invalid chairs and ambulances (Dunlop 1888: 1). But engineers thought that it was unsafe for mainstream bicycles because of the possibility of side-slipping on wet surfaces. More importantly the public believed that air tyres were an ugly feature of the new artefact as can be seen from this comment about the Stanley Exhibition of Cycles (1890: 107): 'the appearance of the tyres destroys the symmetry and graceful appearance of a cycle, and this alone is, we think, sufficient to prevent their coming into general use'.

In the end the air tyre became adopted by bicycle technology largely because of the influence of the social group of sports cyclists, who reinvented its meaning to fit with their need to go as fast as possible and win races. In this sense the initial problems of vibration, side-slipping and ugliness associated with the relevant social groups of engineers and the general public were redefined as a problem of speed according to the interests of sport cyclists. As Pinch and Bijker (1987: 35) note:

> When, for the first time, the tyre was used at the racing track, its entry was hailed with derisive laughter. This was however, quickly silenced by the high speed achieved, and there was only astonishment left when it outpaced all rivals (Croon 1939). Soon handicappers had to give racing cyclists on high-wheelers a considerable start if riders on air-tyre low-wheelers were entered. After a period no racer of any pretensions troubled to compete on anything else.

The important lesson to be learnt from the evolution of the air tyre as a standard feature of bicycle technology is that if another relevant social group such as the engineers or the general public had been more influential than the sports cyclists something else might have happened to the development of bicycle. As a result, a whole industry for the production of air tyres – not only for bicycles but also for cars and airplanes – might not have emerged in the twentieth century. The development path of a new technology is therefore to a large extent socially constructed by one or more relevant social groups who influence the design and stabilisation of technical artefacts and underlying technologies in their early stages of development.

2.5 Technological communities

No matter which social groups participate in the shaping of a new technology it can be argued that there is always an inner circle or core of actors which critically influence its development path. Constant (1980) put forward the notion of a technological community mainly based on the ideas of the historian of science Thomas Kuhn. Kuhn (1970) argued that the cognitive locus of science is a well-defined community of scientists which is associated with some paradigm. Constant (1987) argues that the social locus of technological knowledge is a community of practitioners which creates and follows a technological tradition of practice associated with the evolution of a particular technology. A good example of the existence of technological communities is the different professional engineering societies, such as the Institutes of Electrical and Electronics Engineers (see www.ieee.org/organisations/tab/society.html). These kinds of professional grouping are well institutionalised, highly specialised and well-defined social entities which embody knowledge, development and innovation at the collective level. Such technological communities may be composed of individual adherents, private firms, university laboratories and government organisations. Within these larger communities various interest groups are usually organised focusing on particular problems and technologies for generating knowledge, developing novel applications and setting standards.

Technological traditions of practice bind these communities of technological practitioners together. In this sense technological traditions are analogous to Kuhn's (1970) paradigms and exemplars which scientists share in their invisible colleges (Crane 1972). However, as Constant (1984) argues, it is not clear that a technological tradition of practice comprises a set of specific exemplars in the same sense that a scientific paradigm in its most narrow and precise usage does. A technological tradition has both knowledge and socio-cultural dimensions. The former includes scientific theory and methods and hardware and software, while the latter includes social and communication structures as well as a system of values and beliefs (see Figure 2.2). In other words a technological tradition of practice encompasses a wide array of different elements including scientific theory, explicit and tacit knowledge related to specialised instrumentation, persisting patterns of social interaction and a set of values and beliefs which guide the actions of its members.

There are strict entry requirements into technological communities which usually depend on previous disciplinary training and formal academic

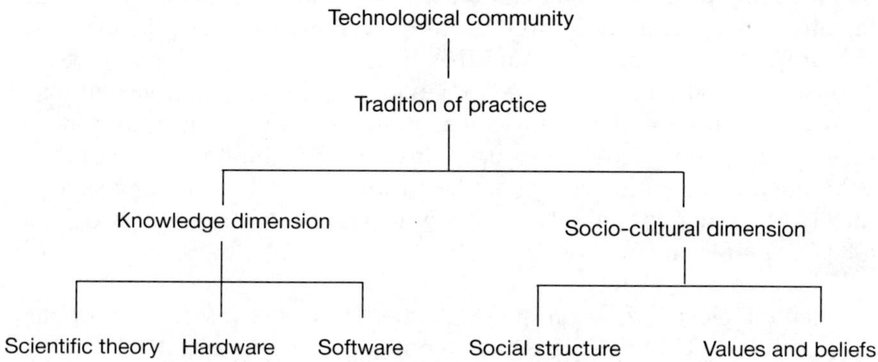

Figure 2.2 Technological community and basic elements of its source concept of technological tradition of practice.

Source: Assimakopoulos 1997a.

qualifications and also on practice and experience gained from senior practitioners through legitimate peripheral participation (Lave and Wenger 1991), socialisation in CoPs within organisational boundaries (Wenger 1998), and participation in NoPs (Brown and Duguid 2001) or occupational communities (Van Maanen and Barley 1984). Disciplinary background is therefore considered a necessary but not a sufficient condition for the rite of passage to full membership in a technological community. An understanding of its socio-cultural dimension, i.e. social structure, as well as its system of values and beliefs, is also required before somebody becomes a full member of this kind of technological community. In the early stages of development of new technological communities persisting patterns of social interactions, a sense of belonging, and shared beliefs and values are more important than disciplinary background and formal academic qualifications because emerging traditions of practice have not fully developed a coherent body of scientific theories and methods as well as formal degrees and qualifications.

The normal task of technological communities is the extension and articulation of the received tradition of practice. Developmental problems for particular technological traditions of practice appear as incapacity to function under new or more stringent conditions, or an inability to maintain performance in a changing world. Normal technology, or what technological communities usually do, involves the improvement of the accepted tradition or its application under new or more stringent conditions. The solutions of normal technology, however, are sought within the limits of the received technological tradition, as both the problems and solutions are presumed to exist within those limits. In a sense community practice defines a cognitive universe that inhibits the recognition of radical alternatives to conventional practice. Members of a technological community usually identify themselves with a particular technological tradition of practice which also ensures social and economic status. As a result inertia and/or the vested interests of any one technological community can create a resistance to change to protect the jobs of its members as well as its power position in existing social structures, such as technical committees (Rosenkopf and Tushman 1998).

Constant (1980) argues that the best evidence for the cognitive and functional dominance of technological communities and their traditions of practice comes in episodes of abrupt transition or technological revolution. Technological revolution can be defined as the professional commitment of a newly emerging community to a new technological tradition. Constant (1980: 20) points out:

> Technological revolution is here defined only in terms of a community of technological practitioners: it has nothing to do with social significance, economic impact, or tawdry advertising. A technological revolution occurs when a community of practitioners embraces a new tradition, whether that community consists of three little old wine makers or of every major aero-engine manufacturer in the world.

From the outset Constant's model of socio-technical change views a handful of people within one or more technological communities as responsible for radical changes in a technological tradition of practice. This view of technological revolution is similar to Kuhn's (1970) and Crane's (1972) models of scientific revolution. Constant (1980: 15) points out that a technological revolution for a tradition of practice is often initiated by presumptive anomalies which arise from science implying that under future conditions a technological system will fail, or function badly, or that a radically different system will do a much better job. See, for example, the case of hypersonic aircraft and underlying technologies where 'at hypersonic speeds beyond Mach 6 … the state of science does not yet permit the development of predictive models and therefore is of limited usefulness in the elaboration of designs and innovation' (Gibbons *et al.* 1994: 20). No functional failure exists in the conventional technological system but anomaly is presumed to exist, hence the expression 'presumptive anomaly'.

Presumptive anomalies are science-based, so individuals who are closely linked to the scientific foundations of their technological tradition can more easily perceive and point out their significance for technological change. Academics doing research on the development of technological systems in their area of interest, or young graduates who are mainly outsiders to their respective tradition of practice but have been exposed to the leading edge of the radical scientific theories and methods, are usually the first actors in a position to understand presumptive anomalies for existing technological traditions of practice. These actors and their aggregates, such as university laboratories and young entrepreneurial firms undertaking RTD work, often drive new technological revolutions, overriding old traditions of practice while creating new opportunities for whole industries and challenging established incumbent firms.

Presumptive anomalies are initially recognised and adopted only by a small minority of members of existing technological communities. Initially people and groups who can understand presumptive anomalies are perceived by the vast majority of an existing technological community as mavericks (Merton 1988), who do not use the accepted scientific theories and methods and do not comply with the norms, values and beliefs of the existing tradition of practice. More importantly, a difficulty with the study of technological revolutions initiated by presumptive anomalies is that after radical change both old and new technological communities and traditions of practice can coexist and co-evolve as they often relate to different tasks. Individuals, organisations and groups forming a new technological community can play multiple roles as they can be members of both old and new technological traditions at the same time. Community commitment need not be total with respect to the development of a new technology.

Constant (1980) successfully points out that two main actors of the car industry, Rolls-Royce and BMW, did not abandon their interests in the car tradition of practice when they became major designers and producers of turbojets for airplanes. The same private companies who played a central role in the creation of the old car technological tradition of practice continued to play a major role in the new turbojet tradition of practice. Multiple membership in different technological communities usually generates a creative tension between different traditions as members of different groups and associated CoPs move ideas from the old tradition to the new one and vice versa. In contrast to the technological revolution, the scientific revolution as viewed by Kuhn (1970) and Crane (1972) implies the rejection of the old paradigm in favour of the new and, in consequence, the abandonment of one system of normal scientific practice and scientific community for another.

Taken together, then, the concepts of technological community, of community of practice, and of the personal 'glocalised' network community, are useful in studying how new knowledge-intensive technology is socially constructed in its early critical stages of development. The concept of personal 'glocalised' community highlights how various actors and social groups related to a new technology, such as GIS, attach meaning to GIS innovations over great geographical distances. The concept of community of practice highlights how people create and share context-specific technology-related knowledge within organisational settings and how they use their NoPs to share explicit and tacit knowledge beyond organisational boundaries. Last but not least, the concept of technological community highlights how a handful of actors who come from various disciplinary backgrounds recognise presumptive anomalies related to a particular domain of knowledge – say, geographic information handling and analysis – and as a result drive radical technological change in existing technological communities and traditions of practice related to this particular knowledge domain.

As a result of multiple memberships, 'non-canonical' practices and presumptive anomalies, new technological communities and traditions of practice come into being. These are collections of people, groups and organisations which share a common interest in a particular technology, e.g. GIS, and play essentially complementary roles with respect to the adoption and implementation of GIS innovations in a broad range of institutional and disciplinary settings. Most of these actors, like members of other technological communities, are committed to a newly emerging GIS tradition of practice while at the same time keeping their disciplinary and professional affiliations to other existing technological communities and traditions of practices, ensuring socio-economic status and societal recognition. Unlike what happens in scientific communities, members of a new technological tradition of practice do not need to abandon their roles within existing traditions of practice and CoPs within organisational boundaries. As a result there is multiple membership and a two-way movement of ideas and experiences from old to new traditions of practice related to a particular domain of knowledge, such as the handling and analysis of geographic information.

Figure 2.3 shows, for example, how a new GIS community comes into being from various existing technological communities and traditions of practice at a national scale. From this it can be seen that a GIS community emerges when people and groups from a number of existing technological

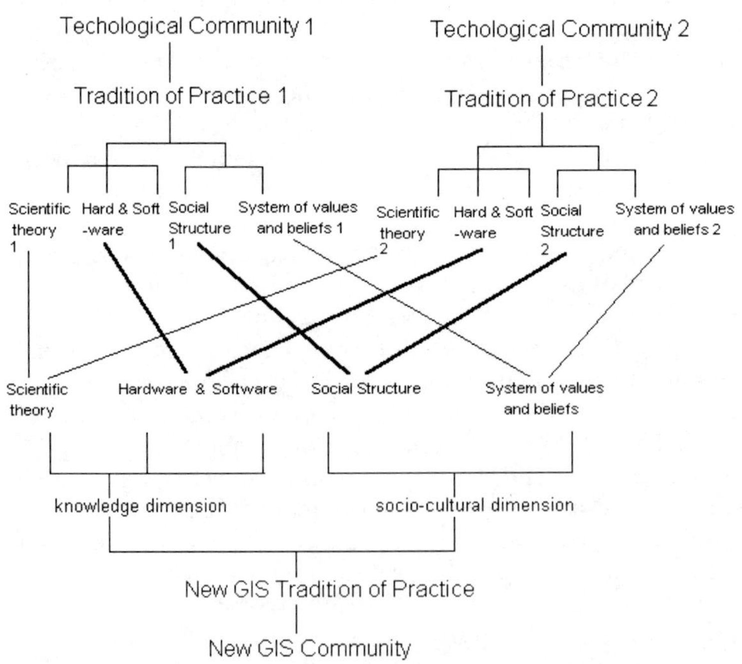

Figure 2.3 Emergence of a new GIS community from existing technological communities and traditions of practice.

Source: Assimakopoulos 1997a.

communities and traditions of practice share and exchange information or other resources to a substantial extent with respect to GIS adoption and implementation. Each one of these existing technological communities (e.g. surveying engineers, computer scientists, urban planners) views GIS technology through its well-established technological tradition of practice. Therefore the new GIS community draws upon various scientific theories and methods, hardware and software as well as on the social and cultural systems of the parent communities. In Chapter 5 the analysis of the emerging Greek GIS community focuses on the social and cultural systems of the parent communities (e.g. surveying engineers, urban planners) with particular emphasis on one element of its knowledge dimension, GIS software, and one element of its socio-cultural dimension, social structure. These socio-technical elements highlight the heterogeneous context within which new technologies are embedded at a national scale and also how different pre-existing technological communities socially construct the development path of a new technology across a broad range of organisational, institutional and disciplinary settings.

2.6 Summary

The first section of this chapter introduced the notion of personal network communities, also highlighting the increasing importance of 'glocalised' communities without propinquity (Wellman 2002). Because of political will and social, economic and technological factors, communities are increasingly organised and operate over great geographical distances. Invisible colleges in science (Crane 1972) are an example of how people from different places have sustained such network communities without propinquity because of common interests for centuries. Moreover the connection and increasing integration of the technical communities of entrepreneurs and venture capitalists in regions of global technology innovation and entrepreneurship, such as Silicon Valley and Hsinchu, were pointed out. The latter is a successful illustration of the value that bring these communities to form the new 'Argonauts' (Saxenian 2006).

The second section of the chapter discussed critically the theoretical perspective of the community of practice as it was developed by Wenger and his collaborators (2002, 1998). It was argued that CoP theory as a social learning theory can shed light on the stickiness of tacit context-specific knowledge related to new technology within organisations, and also explain the leakiness of such knowledge across organisational boundaries through networks of shared practice (Brown and Duguid 2001). Some limitations of the concept of CoP were also identified and discussed. From the outset it seems that the concept of CoP is rather amorphous, group-centric and inward-looking. Today much knowledge-intensive technological development work in critical application areas (i.e. ICT, biotechnology) for the future takes place across the organisational boundary through all sorts of inter-organisational alliances and inter-personal networks. Moreover CoP theory seems to downplay the centrality and power of some actors and relevant social groups in the shaping of distributed complex technologies. The latter limitation is properly addressed within models of socio-technical change stemming from the field of STS.

The third section of the chapter therefore focused on the science–technology relationship, arguing that science and technology have become intermixed over the past few decades and that there is now increased interaction between various actors who come from different institutional and disciplinary settings (Gibbons *et al.* 1994; Etzkowitz and Leydesdorff 2000). Academics, government officials and private sector managers increasingly collaborate or/and compete for the same resources at a local, national and international scale, trying to define what a new technology can and cannot do. Two socio-technical change models were therefore presented to shed light on how different actors and groups try to define and shape new technologies. These are the social constructivistic model developed by Pinch and Bijker (1987) and the concept of technological community put forward by Constant (1980, 1984, 1987). Pinch and Bijker include in their analysis a wide range of social groups, as well as engineers and practitioners, highlighting the influence of relevant social groups in the shaping of new technological systems.

Constant's model of socio-technical change is based on Kuhn's (1970) argument that the social locus of scientific knowledge is a well-defined community of scientists that shares a paradigm or exemplar. Constant argues that the social locus of technological knowledge is a well-defined community of technological practitioners that creates and follows a technological tradition of practice. This tradition of practice encompasses both knowledge and socio-cultural dimensions. In this sense a tradition of practice is the glue of a technological community. Developmental problems appear in technological communities because members cannot sustain improved performance in a changing world. As a result various members who are close to the scientific foundations of their particular tradition of practice recognise presumptive anomalies and initiate radical technological change or technological revolutions which override old traditions of practice. As a result of such 'non-canonical' innovative practice a new technological community emerges together with a new technological tradition of practice. This model of socio-technical change was applied with respect to GIS technology to illustrate the emergence of a new technological community and tradition of practice as well as the value of multiple memberships in both old and new technological communities and traditions of practice.

3 Collaboration networks as the social locus of knowledge-intensive technological innovation

3.1 Introduction

This chapter starts from the processes of technological innovation rather than the communities who socially shape new ideas and technologies. It highlights at multiple levels of analysis the increasingly important phenomenon of collaboration and networking as the social locus of knowledge-intensive technological innovation. Section 3.2 introduces recent non-linear innovation models fostering collaboration and networking among firms, universities and government institutes. The emphasis of the discussion is kept on speed and flexibility of RTD work, highlighting also the shift from 'mode 1' to 'mode 2' knowledge production (Gibbons *et al.* 1994), and the triple helix model of university–industry–government relations (Etzkowitz and Leydesdorff 2000). Section 3.3 focuses on European innovation policies supporting collaboration in ICT RTD across organisational, institutional and national boundaries via new forms of network organisations such as collaborative projects. For example, the European Strategic Programme for Research in Information Technologies (Esprit) subsidised more than 2000 RTD projects linking a minimum of two partner organisations from a minimum of two EU member states from the early 1980s until the late 1990s.

Section 3.4 shifts the focus of the discussion from government policy to firm strategy. The discussion highlights the importance of distributed technological capabilities in innovation, and that emerging technologies are now less frequently produced and located within single firms and increasingly distributed across a range of firms and their suppliers, customers and other knowledge-generating institutions such as universities. Section 3.5 discusses the importance of personal networks for RTD work and innovation, especially in high-velocity markets such as ICT. Particular attention is paid in this section to informal networks of a social nature linking people and firms, facilitating the flow of knowledge and exchange of information for innovation. Finally Section 3.6 is the chapter summary.

3.2 Non-linear models in knowledge-intensive innovation

Technological innovation is a concept with a long history in economics and other disciplines such as management and science and technology policy. Scholars in the economics of innovation literature (see, for example, Rothwell 1992; Dodgson 2000; Pavitt 2003) have put forward various models of the innovation process that challenge earlier views of technological innovation as a linear process of technology push or market pull. According to Rothwell's (1992) review of studies and models of innovation processes since the 1950s, there are five generations of models describing innovation processes (see Figure 3.1). Figure 3.1 shows that early innovation models conceptualised innovation as a linear process putting the emphasis either on RTD or marketing, or a sequential coupling process of RTD and marketing with feedback loops. More recent models that have emerged in the past two decades or so, i.e. the fourth and fifth generations, put a premium on RTD collaboration, external networking and speed to market for determining a company's competitiveness, especially in knowledge-intensive areas, such

FIRST GENERATION

Technology push. Simple linear sequential process. Emphasis on R & D. The market is a receptacle for the fruits of R & D.

SECOND GENERATION

Need pull. Simple linear sequential process. Emphasis on marketing. The market is the source of ideas for directing R & D. R & D has a reactive role.

THIRD GENERATION

Coupling model. Sequential, but with feedback loops. Push or pull or push—pull combinations. R & D and marketing more in balance. Emphasis on integration at the R & D/marketing interface.

FOURTH GENERATION

Integrated model. Parallel development with integrated development teams. Strong upstream supplier linkages. Close coupling with leading-edge customer. Emphasis on integration between R & D and manufacturing (design for makeability). Horizontal collaboration (joint ventures, etc.).

FIFTH GENERATION

Systems integration and networking model (SIN). Fully integrated parallel development. Use of expert systems and simulation modelling in R & D. Strong linkages with leading-edge customers ('customer focus' at the fore-front of strategy). Strategic integration with primary suppliers including co-development of new products and linked CAD systems. Horizontal linkages, including joint ventures, collaborative research groupings, collaborative marketing arrangements, etc. Emphasis on corporate flexibility and speed of development (time-based strategy). Increased focus on quality and other non-price factors.

Figure 3.1 Five generations of models describing innovation processes.

Source: Dodgson 2000: 42, based on Rothwell 1992.

as ICT where rates of technological change are high and product cycles are short.

In particular, the fifth-generation model of systems integration and networking highlights the increasing significance of time-based technology strategies and collaboration in competition for surviving in what Hindle (2003: 107–8) calls the third age of globalisation today. After marketing (first age) and manufacturing (second age) successive generations of knowledge-intensive products and services globally, RTD work has increasingly been outsourced and carried out in a distributed global scale for shortening lead times and sustaining continuous innovation (Arora and Gambardella 1990; Howells 1999; Saxenian 1990, 2002; Saxenian and Hsu 2001). For example, the most innovative multinational companies in the ICT industry, such as Intel, have benefited from manufacturing successive generations of their increasingly sophisticated products outside the United States, in Taiwan and India. Moreover, they have recently started moving RTD work nearer to their key markets and customers in Asia. Customer focus, strategic integration with primary suppliers and horizontal linkages, such as collaborative groupings and linkages with competitors, fuel collaboration and networking for continuous and time-based innovation. Mass customisation, often making use of emerging Internet technologies, has also empowered lead customers and user communities to get involved and play a defining role in innovation processes of such diverse products as semiconductor chips, open source software and sports goods (Von Hippel 2001; McKelvey 2001).

Today the underlying reasons for networking and collaboration, including the externalisation and outsourcing of RTD work, are manifold. Research and development work has become increasingly complex and uncertain worldwide, product and process innovation life cycles get shorter, while RTD work requires more knowledge resources and take longer, transcending more and more scientific and technological communities. The reduction of lead times and the expansion of the knowledge base of participants through complementary talents and capabilities, better market positioning, as well as the sharing of RTD costs and risks which are scaling up in an exponential manner, are some of the key benefits of collaboration and networking in knowledge-intensive and high-velocity markets such as ICT (Hagedoorn and Schackenraad 1993; Coombs *et al.* 1996; Assimakopoulos 2003). As is discussed in more detail in the next section, government innovation policies supported by the EU and other national governments have also fuelled the growth of the phenomenon of collaboration networks throughout Europe and beyond since the early 1980s (Mytelka and Delapierre 1987; Charles and Howells 1992; Georghiou 1999).

Two recent non-linear innovation models, focusing on the dynamics of science and research in contemporary societies (Gibbons *et al.* 1994) and university–industry–government relations (Etzkowitz and Leydesdorff 2000), are worth introducing next. Gibbons *et al.* (1994) discuss the new production of knowledge in contemporary societies and economies through what they call transformation from 'mode 1' to 'mode 2' knowledge production. Mode 1 refers to the traditional disciplinary cognitive and social structures of science and technology, where knowledge is produced within an invisible college (see Section 2.4) or technological community (see Section 2.5) and according to a certain paradigm or tradition of technological practice. In contrast to 'mode 1' the new 'mode 2' knowledge is produced within the context of application (e.g. GIS) and it is better described as applied rather than basic research and development. According to Gibbons *et al.* (1994: 3–8) 'mode 2' knowledge production is characterised by some if not all of the following characteristics: trans-disciplinarity, heterogeneity and organisational diversity, social accountability and reflexivity, and quality control.

'Mode 1' knowledge is often produced in universities, while 'mode 2' knowledge production most often involves a broad range of actors and collaboration networks from academia, as well as private firms and government agencies. Generally speaking 'mode 2' does not supersede 'mode 1'. 'Mode 2' is an outgrowth of 'mode 1' in the manner illustrated with regard to GIS technology in Section 2.5 (see also Figure 2.3). The new 'mode 2' is also complementary to 'mode 1' in the sense that it is 'accumulated through the repeated configuration of human resources in flexible, essentially transient forms of organisation' (Gibbons *et al.* 1994: 9), such as 'project networks' among collaborating firms, and other actors such as universities, within a specific time frame (see Chapter 4). From the outset, it seems that the new 'mode 2' has emerged alongside the traditional disciplinary structure of science and technology and 'mode 1' through the emergence of a socially distributed knowledge-production system, such as the one described by the triple helix model of the university–industry–government relationship put forward by Etzkowitz and Leydesdorff (2000).

According to the triple helix model, RTD work is no longer organised into separate non-intersecting institutional spheres of academia, industry and government, but the different spheres increasingly interact and cross over on an equal basis. Etzkowitz and Leydesdorff (2000: 113) argue that:

> The various negotiations among the spheres operate on a roughly equal basis. Industry and government have traditionally been the leading institutions in advanced industrial societies and academia is moving up into this category, as more activities in both government and industry become dependent upon advances in knowledge. There is also more cross over and cooperation within as well as across institutional spheres, given both the increasing complexity of tasks and the rapidity of technical advance, for example, in generations of computer chips. Both RTD and marketing move into networks because individual companies can no longer perform all the work, even to bring many products to the market. So they have to form partnerships within industry. Sometimes the research they need is longer term so that even if a government

program just focuses on funding companies, the companies find that they have to give subcontracts to university researchers.

In the last couple of decades the study of university–industry relations in particular has provided insights into the strengthening of collaboration networks among academics based in university research centres and private firms that not long ago used to interact across strongly defended boundaries. Today universities are seen as key partners in knowledge-intensive innovation networks, as besides their traditional dual mission of teaching and academic research, a third mission of linking up with industry has been added to their portfolio of activities, with science and technology parks, business incubators and the like gaining increased prominence on or near university campuses worldwide. Academics from the engineering faculties of leading US universities such as MIT and Stanford have long been actively encouraged to embrace a third mission besides their academic duties and either undertake entrepreneurial roles in industry – for example, start companies where the university has a stake, join the technical or management boards of young or established companies that the university has an interest in (Kenney and Goe 2004) – or/and carry out contract research externally funded by government agencies, such as the US Department of Defense (Leslie 2000). For example, in the past couple of decades Stanford University faculty or/and students have played key roles in establishing such entrepreneurial companies as SUN Microsystems, Netscape and Google.

However, experiences vary widely across the Atlantic and UK universities are considered more entrepreneurial than their counterparts across the European continent. For example, Howells and Nedeva (2003) argue that today research-led universities in the UK can easily transcend the national system of innovation and take advantage of opportunities for carrying out research either funded by the EU Framework programmes or multinational companies that externalise and sub-contract RTD at a global scale. On the other hand, German universities still find themselves in the shadow of von Humbolt, the founder of the first technical university in Berlin almost two centuries ago, who believed in a socially disembedded and autonomous university, where theoretical advancements are widely recognised from academics and students alike, but empirical research and industrial applications are significantly lower in status (Krucken 2003).

Despite the significant differences in institutional environments across both sides of the Atlantic, as well as the well-documented argument that the Silicon Valley model has grown organically without much government planning or intervention (see, for example, Macdonald 1983; Rogers and Larsen 1984; Saxenian 1990, 1996; Rowen 2000; Kenney 2000), it is worth examining in more detail the role of government innovation policies, at least in Europe, in conjunction with ICT research, since these have fostered collaboration and networking among ICT firms, universities and other related knowledge-generating institutions throughout the EU.

3.3 European policies fostering collaboration in ICT RTD

In the early 1980s European firms began to realise that their industrial and technology-related capabilities were lagging behind in such core high-technology areas as ICT and some had already begun to collaborate (Mytelka and Delapierre 1987). Policy makers were also becoming increasingly concerned about the gradual loss of competitiveness they perceived in the European economy and in the European ICT industry in particular. The globalisation of high-technology industries (Narula 1999) and the wide disparities between the industrial and technological capabilities of the various country members revealed by the continuing expansion of the EU (especially evident in the divide between the wealthy countries of the European north and the poor countries of the European south) further reinforced this perception (Hagedoorn *et al.* 2000). Moreover, policy makers on both sides of the Atlantic had become very enthusiastic about 'Japanese-style' collaborative research and the perceived success of 'keiretsu' (Ray 1998; Georghiou 1999).

European industry generally was beginning to show much more interest in collaborating in RTD, previously an activity conducted secretly and independently of competitors' research and development efforts (Narula and Hagedoorn 1999). According to Narula (1999), the underlying objective of the Framework programmes of the EC was not to encourage collaboration per se; rather it was to encourage collaboration in the run-up to the single European market in 1992. Collaboration would allow EU industry to restructure and be better able to face the competitive environment of the single market. It was hardly surprising, then, that collaborative RTD became central to the EC's policy in the early 1980s (Peterson 1991), and thus that collaboration became central to its first and most significant RTD programme, Esprit.

Before Esprit is introduced, however, it is important to bear in mind the era of which Esprit was so very much a product. By the early 1980s the importance of what had come to be called the 'microelectronics revolution' was evident, and not just in terms of the technological innovation that helped make the electronics firms themselves more competitive (Braun and Macdonald 1982). The application of the transistor and then the integrated circuit (see also Chapter 6) in computing and in telecommunications had begun to transform products and processes in virtually all industries, and consequently the competitiveness of all was increasingly seen to be dependent on the new ICT-related technologies, in particular microelectronics. Microelectronics, then, could not help but command political attention.

Governments in most developed countries were compelled to calm fears about the disruption the new technology would cause. This presented a minor challenge, but where there is challenge there is also opportunity and microelectronics offered a major opportunity of the sort that is attractive to politicians and policy makers. By the early 1980s it was common practice to perceive microelectronics as but one – although certainly a major one – of a series of new and related technologies considered collectively as 'high technology'. The benefits of high technology were reckoned to be immense, far outweighing its costs. So even if some jobs were lost and others de-skilled, new jobs would be created and of a very superior sort. And if high technology sounded the death knell of the old smokestack industries, it brought shiny new ones in their place. The political advantage of being able to bring about such social and economic benefits was not lost on governments, and everywhere they sought to become involved with high technology and microelectronics in particular. Their recent determination to be associated with Internet technologies, such as electronic commerce is not dissimilar (e-Europe 2002; Cabinet Office 1999).

There was, of course, no doubting the competitive advantage that US and Japanese firms held in microelectronics. European governments might console themselves with the appreciation that the advantage of the former had come about through the country's head start (see Chapter 6), and that they might erode this lead by replicating in Europe the circumstances under which high technology flourished in the United States – a case of policy making up for the European market's failure to act as the US market had done. But the Japanese case offered no possibility of such consolation; there was certainly no way that a Japanese environment could be re-created in Europe. Or was there? European policy makers allowed themselves to become convinced that the Japanese had become so successful in microelectronics because MITI, the Ministry of International Trade and Industry, had engineered the coordination of government, industry and universities to be innovative in microelectronics.

Indeed, when Esprit was launched, the European Commission had grand ambitions to be a European MITI (Dickson 1983). The Fifth-Generation Computer Programme was trumpeted by the Japanese to European governments already convinced that innovation in microelectronics came through collaboration and that collaboration could be arranged by government (Newman 1982). That Japanese collaboration and coordination was as much a product of Japanese culture as of the efforts of MITI, and that this collaboration and coordination was deeply dependent on personal links and obligations, was conveniently overlooked, much as the informal and mostly privately supported innovation of Silicon Valley was disregarded (Galinski 1984; House of Commons 1988).

So Europe, and especially the EC, imported from Japan the notion that formal collaboration arranged by government was essential for ICT-related innovation and hence for competitiveness in microelectronics, and imported from the US evidence of the conditions required for high-technology industry. It was clear that government in Europe had to do something for microelectronics, and equally clear that misconstruing both the Japanese and the US model permitted considerable latitude (Mackintosh 1979). For example, national champions – the firms that dominated the European electronics industry and that were to dominate the early Esprit – could be portrayed not simply as socially and economically integrated in the Japanese fashion, but also as enterprising and entrepreneurial in the Californian way. The collaboration of Esprit might seem very different from the personal information networks of Silicon Valley, but in the context of the scramble by governments everywhere for association with high technology, and in the absence of the need for any appreciation of how high technology really works, formal collaboration in pre-competitive research was quite acceptable.

In 1981, the EC suggested that European ICT firms, including national champions such as Siemens, Philips, Thomson and Olivetti, take a concerted approach to ICT research, and invited their collaboration in drawing up a common strategy (Mytelka and Delapierre 1987). Following the launch of a small pilot programme in 1983, Esprit proper was started in 1984. There have been four phases of Esprit research (Esprit I, 1984–7; Esprit II, 1987–90; Esprit III, 1990–4; and Esprit IV, 1994–8) – all jointly funded by the EC and the participating organisations (Assimakopoulos and Macdonald 1999). The fifth Framework programme (1998–2002) initiated the Information Society Technology (IST) programme, placing all EC ICT research under one umbrella programme, which continues as a thematic priority for the current sixth Framework programme (2002–6).

The early Esprit was very much driven by the belief that collaboration and networking among industry, universities and public research institutes across Europe was an effective means of narrowing what was perceived as a technological gap between European companies and their American and Japanese competitors (Mytelka and Delapierre 1987; Gulati 1995; Mytelka 1995; Narula 1999; Hagedoorn *et al.* 2000; Piekkari, Macdonald and Assimakopoulos 2001). As Mytelka and Delapierre (1987: 233) point out, collaboration among European firms was more attractive than alliances with non-European firms because it was thought to involve less risk and to enable firms to take advantage of economies of scale in one or more of their production processes while remaining separate industrial entities.

The research consortium – termed the 'project' by the EC – has long been the primary unit of Esprit and IST organisation. The project has often seemed to be the only unit, though it is worth noting that new instruments for funding in the sixth Framework programme, such as integrated projects, have allowed the networking and clustering of a set of projects with a similar set of objectives for generating synergies and adding value to individual projects and to key actions or priorities at the programme level. Nonetheless all EC and Esprit organisation has been centred on the project for most of the past two decades, as has most monitoring and evaluation (Georghiou 2001). In 16 years (1983–98), some 2250 Esprit projects have been completed and more than €5.5 billion has been spent (Assimakopoulos and Macdonald 1999).

The project officer – the key EC official – tended to regard projects as self-contained, to be completed within a specific timeframe as specified by a formal contractual agreement between the EC and the main contractor of each and every project consortium. In the 1990s, Esprit has been through vast changes in its organisation and scope (Assimakopoulos *et al.* 2000). The EC has responded to new trends in the collaborative behaviour of the IT industry by, for example, expanding Esprit participation, encouraging collaboration throughout the IT value chain, and increasing emphasis on the users of IT as well as SMEs. Some of these developments are summarised in Table 3.1.

Despite these alterations in emphasis, many of the characteristics of the early Esprit were evident until the conclusion of the programme in 1999. For example, Esprit always insisted that the research it supported be collaborative in nature, specifically that there had to be a minimum collaboration in each project of two partner organisations from two EU member countries. The early Esprit was also determinedly pre-competitive, focusing on research that was considered to be distant from the individual market interests of collaborators. The notion of pre-competitive research provided a convenient label for the activity undertaken within collaboration, one acceptable to the free market ideology of most European governments of the period (Georghiou 1999). It was argued that collaboration in pre-competitive research did not constitute government interference in market forces (Quintas and Guy 1995) and fitted comfortably within a technology push model of ICT innovation.

Sweeping changes in the ICT industry, together with improved understanding of how innovation is generated, however, encouraged Esprit to change its emphasis from technology push in the 1980s to market pull in the 1990s during the third and fourth Framework programmes. This has required

Table 3.1 Summary of changes in Esprit and IST programmes from the early 1980s to the early 2000s

Dimension	*Esprit (1983–1998)*	*IST (1998–2006)*
Participants in collaboration	Dominance of electronic firms, IT suppliers, and participants from northern Europe as well as less favoured regions	A heterogeneous group of organisations representing the entire IT value chain and including SMEs and user organisations
Nature of collaboration	Pre-competitive	Collaboration in competition
Focus of collaboration	Hard science	Soft science (emphasis on socio-economic research)
Organisation of collaboration	Research project	Research clusters and networks
Role in the broader community	Inward-oriented, isolated	Outward-oriented, integrated

Source: Assimakopoulos *et al.* 2004: 249.

abandoning the idea that partners can collaborate only when they are being pre-competitive. It has been accepted that they may also collaborate when they are cooperating in competition (Hamel, Doz and Prahalad 1989). Indeed, the success of the IST programme is dependent on the willingness and ability of partners to collaborate in competitive circumstances. The early Esprit was dominated by the rigid conviction that innovation emanated, quite obviously, from science and engineering. Just as the model of innovation within Esprit has changed from technology push to market pull, Esprit and IST research is no longer confined to science and engineering and now includes at least some social science research by explicitly requiring in the sixth Framework programme that each and every project articulate what is the contribution it makes in terms of socio-economic issues.

Today the IST programme therefore acknowledges that socio-economic research cannot be isolated to a single domain, such as the task 0.9 'analysis of socio-economic consequences' in Framework Programme IV (Assimakopoulos *et al.* 2000), or the key action 2.1 'socio-economic research' in Framework Programme V, but must underpin all its ICT research. In consequence, the IST programme in Framework Programme VI cannot be accused of fostering innovation intended to benefit only the suppliers of ICT equipment: IST innovation is now directed towards all users of ICT including small firms, universities, etc. It has been accepted that there is now no part of the economy which is not heavily dependent on ICT and IST underlies the whole fabric of the emerging information society and knowledge economy throughout the EU and beyond (Castells 2001).

The changes that Esprit has therefore undergone in terms of participation, focus, organisation and orientation were responses to particular trends and developments in the ICT industry, and more general shifts in the competitive environment of deepening globalisation in marketing, manufacturing and RTD (Hindle 2003). Throughout the history of Esprit, the main policy objective of the EC has been to create and sustain a fertile platform for collaboration, innovation and networking. However, it is difficult, perhaps impossible, to confine collaboration and networking, and to harness innovation by restricting them to a single geographical region, even one with all the resources of Europe. More importantly it may be pointless, as will be discussed in Chapter 4.

3.4 Firms' strategies and distributed innovation capabilities

Since the 1980s, institutional economists and knowledge management scholars have pointed out that even firms as large as IBM cannot marshal independently all required resources and capabilities for time-based innovation (Winter 1987; Badaracco 1991; Boisot 1998; Coombs and Metcalfe 1998; Coombs *et al.* 2001; Teece 2001). Large and small firms alike have to rethink their 'collaboration' strategies and decide whether they can create and sustain competitive advantage on their own, or whether they have to collaborate within dyads, or networks of partners, for generating and acquiring the necessary knowledge and external capabilities from other rival firms (Von Hippel 2005, 1988), as well as non-rival knowledge-generating institutions such as universities. According to Coombs and Metcalfe (1998) the increasing importance of external capability acquisition for fostering technological innovation in the early 2000s is the result of a complex strategic process of creating, combining and exploiting knowledge and capabilities in Richardson's (1972) territory of cooperation between firms.

Coombs and Metcalfe (1998) highlight the significance of distributed technological capabilities within a network of collaborating and/or competing firms as the key for successful knowledge-intensive innovation strategies today. Moreover, their discussion about distributed technological capabilities illustrates the technological diversity and systemic complexity of innovations today, as well as the increasing connectedness of science and technology knowledge bases. As Kodama (1992) has persuasively argued, technological innovations for the past decade or so embody an increasing range of technologies. 'Technology fusion', or combining emerging technologies into hybrid technologies, has opened up a broad range of opportunities for new RTD work – for example, in bioinformatics – shifting away the focus from 'breakthrough' innovation according to the old maxim 'one technology – one industry'. Technological innovations are also embedded in a context of use which is becoming increasingly systemic in that new products and processes need to inter-operate and ensure compatibility with other products and processes in their environment (Preece and Laurila 2003). Moreover, technological innovations stem from an increasingly wider range of scientific and technological practices and connected knowledge bases as described in 'mode 2' knowledge production and the triple helix model of university industry–government relationships above and illustrated in the context of GIS in the previous chapter (see Figure 2.3).

For the past two decades it has therefore been increasingly recognised by academics and private sector managers alike that technological innovations cannot be accomplished by individual firms acting independently. Firms' strategies have implemented a plethora of networking and collaborative arrangements, such as joint ventures and strategic alliances, research consortia jointly sponsored by firms and government agencies (e.g. Esprit and IST projects), technology exchange and licensing agreements, and subcontracting and production sharing, to name but a few. These new forms of 'network' organisations exist between traditional markets and hierarchies (Thorelli 1986; Freeman 1991; Foss 2002), and differ from traditional forms of organisation in the sense that they are governed by neither market transactions nor organisational hierarchies. They are characterised by varying degrees of control or autonomy, from formal contractual agreements, as in the Esprit project networks or joint venture companies, to informal networks of information and knowledge exchange, presented in more detail in the next section. Overall such new forms of organisation arrange and govern

RTD work in a flexible and distributed manner, fostering collaboration, learning and innovation of the sort highlighted above in the fifth-generation innovation models and EC policy with regard to ICT RTD.

Reflecting on the emergence and growing significance of new forms of organisation for innovation and competition, Dyer and Singh (1998: 660) put forward a 'relational view' of strategy and competitive advantage, arguing that 'the (dis)advantages of an individual firm are often linked to the (dis)advantages of the network of relationships in which the firm is embedded'. This relational view of how firms create, sustain, or lose competitive advantage reflects a 'third way' for conceptualising strategy in the early 2000s, on top of the 'industry structure view' mainly associated with Porter (1980), and the 'resource-based view' put forward by Penrose and developed in the last decades by scholars such as Teece (1987), Barney (1991) and Spender (1996). From the outset, the 'relational view' of competitive advantage shifts the emphasis of the discussion from individual firms as the unit of analysis to the dyadic relationships and inter-organisational networks that a firm's critical resources and capabilities may stem from and depend on.

Moreover the 'relational view' builds on the 'resource-based view' of the firm, with respect to the importance of knowledge underlying inimitable resources and dynamic capabilities, such as new product development and decisions regarding strategic alliances (Kogut and Zander 1992, 1996). Recent insights from business strategy (see, for example, Galunic and Eisenhardt 2001; Eisenhardt and Martin 2000) have pointed out the significance of dynamic capabilities in high-velocity markets such as ICT, where the emphasis of the resource-based view on long-term advantage is often problematic, and short-term unpredictable advantage is the norm. Table 3.2 shows that in high-velocity markets dynamic capabilities such as product development and strategic decision making with respect to networking and collaboration are simple, highly experiential and fragile processes with often unpredictable outcomes. Ambiguous industry structure, porous boundaries and fluid business models are only some of the contrasting differences of dynamic capabilities in high-velocity markets compared to the ones in moderately dynamic markets that are more detailed, relatively structured analytic routines that rely extensively on existing knowledge and stable processes with predictable outcomes (see Table 3.2).

A 'relational view' of strategy and the distributed nature of technological and dynamic capabilities necessary for learning and innovation in high-velocity markets shift the emphasis of the discussion from individual firms to the architecture of collaboration in dyads, triads and networks of collaborating and competing organisations, individuals and their communities of practice (CoPs). A 'relational view' is missing from the CoP theory of learning and innovation in organisations (see Section 2.3, where it was highlighted that CoPs seem to be rather inward-looking, group-centric and bounded within organisational boundaries). Since CoP theory conceptual-

Table 3.2 Dynamic capabilities in moderately dynamic and high-velocity markets

	Moderately dynamic markets	*High-velocity markets*
Market definition	Stable industry structure, defined boundaries, clear business models, identifiable players, linear and predictable change	Ambiguous industry structure, blurred boundaries, fluid business models, ambiguous and shifting players, nonlinear and unpredictable change
Pattern	Detailed, analytic routines that rely extensively on existing knowledge	Simple, experiential routines that rely on newly created knowledge specific to the situation
Execution	Linear	Iterative
Stable	Yes	No
Outcomes	Predictable	Unpredictable
Key to effective evolution	Frequent, nearby variation	Carefully managed selection

Source: Eisenhardt and Martin 2000: 1115.

ises organisations as a constellation of CoPs, it overlooks much of the ongoing discussion on the strategic significance of new forms of organisations for learning, innovation and the acquisition of technological capabilities distributed among a network of collaborating and competing firms and other knowledge-generating organisations such as universities.

This shortcoming of CoP theory with regard to how information and knowledge flows across the firm boundaries has recently been highlighted by Brown and Duguid (2001) in their discussion of networks of practice, or how knowledge flows across the rails of common practice: for example, between private and public organisations in Silicon Valley (Brown and Duguid 2000b). It is also explored in more depth in the next section and Chapter 6 with respect to the semiconductor community in SV.

3.5 Personal and social networks in knowledge-intensive innovation

Silicon Valley is perhaps the geographical context in which the importance of personal and social networks in technological innovation is most recognised and appreciated (Saxenian 1990 and 1996; Castilla *et al.* 2000; Castilla 2003; Ferrary 2003). It is widely accepted that firms both cooperate and compete in this environment, and that social networks enable the speed and extent of innovation the market requires while making possible yet faster technological innovation (Henton 2000). Moreover, it is accepted that much of this networking is informal, involving personal links, and that informal networks are particularly efficient in fostering the exchange of information and knowledge required for rapid and radical innovation locally, but also across great geographical distances (Rogers 1982; Saxenian 2005). From the outset, such social networks reveal how economic action within and across organisational boundaries is structurally embedded in the social networks of individual actors, enabling economic outcomes (Granovetter 1985, 2005).

Sociologists, economists and management scholars among others have put considerable effort into explaining how organisations, and in particular key individuals within organisations, play crucial roles for technological innovation and build social networks and alliances based on their previous direct and indirect ties. In a seminal article, Freeman (1991: 499) introduces the term 'networks of innovators' to highlight the 'vital importance of external information networks and of collaboration with users during the development of new products and processes'. Conway (1995, 1997) also reports that the mobilisation of informal boundary-spanning contacts and networks accounts for between one-third and two-thirds of the inputs important to the development of 35 successful innovations across a broad range of knowledge-intensive sectors in the UK. More recently, Soh and Roberts (2003) report how networks of innovators drove the evolution of complex technological innovations in the data communication industry in the USA from the mid-1980s to the mid-1990s.

Cross and Prusak (2002: 106) have identified four common role-players in informal networks: central connectors, boundary spanners, information brokers and peripheral specialists, whose performance is critical to the productivity and innovation capacity of any organisation. Central connectors link most people in an informal network; boundary spanners connect an informal network with other parts of the company or with similar networks in other organisations; information brokers keep the different subgroups in an informal network together; and peripheral specialists are consulted by different members of an informal network because of their knowledge and expertise in particular areas of interest. Managers, however, do not often recognise these role-players, as they only know their own personal networks (and not those of all their employees) and as a result they often deprive such informal networks and particular role-players of resources and managerial attention. Note that informal networks by their very nature are not part of the official hierarchy (Krackhardt and Hanson 1993) and as Freeman (1991: 502) points out, 'informal networks are extremely important, but very hard to classify and measure'.

In practice, however, Gulati and Gargiulo (1999) argue that organisational decision makers, facing the uncertainty that usually stems from the scarcity of information about the true capabilities and expectations of potential partners in collaboration networks, rely on their personal networks of past partnerships to guide their future alliance decisions. Thus the creation of new ties was often heavily dependent on direct contact in the past with individuals deemed trustworthy (Dodgson 1993). In this sense, personal informal networks play a critical role in the development of new formal inter-organisational networks, enhancing the capacity of the latter to shape subsequent alliance decisions. Burt (2004) has also highlighted the significance of personal informal ties in providing organisations with not only trustworthy 'first-hand' information, but also access to timely information in relation to new projects, and referrals to other organisations in a potential collaboration. Where there is particular uncertainty about potential partners because of geographical, language and cultural barriers, there is an even greater need for a common history of partners and indirect ties to common third parties. This system also deters opportunism, as any bad behaviour from any partner is easily reported to other common partners Gargiulo (1993).

Structural holes and weak ties (Burt 1992, 2004; Granovetter 1973, 1982) have been also highlighted as critical characteristics of social networks for building comparative advantage in an increasingly turbulent and competitive marketplace. Structural hole theory suggests that actors who occupy brokerage positions between cohesive groups enjoy comparative advantages in negotiating relationships which allow them to know about more opportunities and to secure more favourable terms in the opportunities they choose to pursue. Weak ties theory has put forward a similar argument, pointing out the importance of relationships between actors who are embedded in separate but cohesive groups. For example, if the weak tie between actors A and B is removed, then there is no possibility of information exchange between the two teams in the network shown in Figure 3.2. Note that in contrast to the weak tie A–B all other ties in Figure 3.2 are strong in that information can flow between actors through a number of both direct and indirect paths.

Weak ties can link separate individuals or organisations and their associated communities of practice, thereby creating whole new technological communities (Assimakopoulos 2000; Assimakopoulos and Yan 2006b). Low-density networks of weak ties, rather than strong ties, often have a significant value in terms of greater information diversity and as a result value for innovation (Burt 2005). In larger social systems, such as entire technological communities, weak ties may also play key bridging roles that give birth to greater integration. This observation is similar to Blau's (1974: 623) broader sociological argument that:

> since intimate relations tend to be confined to small and closed social circles … they fragment society into small groups. The integration of these groups in society depends on people's weak ties, not their strong ones, because weak ties extend beyond intimate circles and establish the inter-group connections on which macro-social integration rests.

Of course, the distinction between strong and weak ties is often unclear: what is weak and what is strong is not only relative but also dependent on context (Marsden and Campbell 1984; Marsden 1990). As was discussed in the methodology section (see Chapter 1 and Appendix) SNA uncovers and

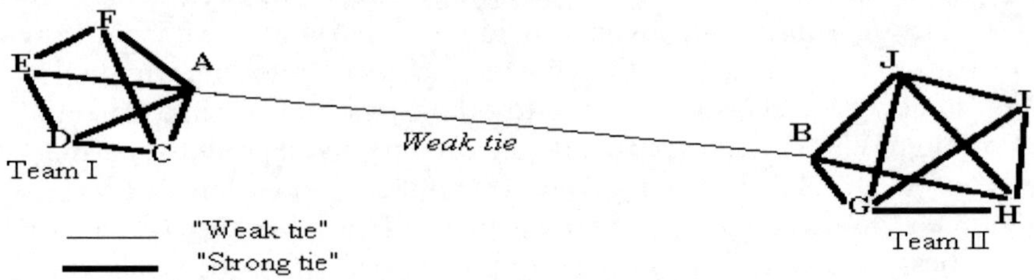

Figure 3.2 A weak tie between the actors A and B.

Source: Rogers 1983: 42.

maps the entire structure of a personal or/and community network and uses different metrics to identify particular roles and positions that actors occupy in it. In fact there is an ongoing debate about how, for example, to measure the 'centrality' (Freeman, Borgatti and White 1991; Bonacich 1987) of actors in a network, or accrue value from 'social capital' (Lin 2002) that actors gain by participating in various networks and communities.

Burt (1992: 12) defines social capital as both the resources personal contacts hold and the structure of contacts in a personal network. According to Coleman (1988) social capital is based on high trust in tightly knit 'cohesive' networks where strong overlapping ties foster mutual assistance obligations and the social control of deviant behaviours (Gargiulo 1993). Burt (1992, 2004, 2005), however, argues that competitive advantages and higher returns for an actor are accrued by weak rather than strong 'cohesive' ties where direct tie relationships span structural holes, therefore gaining earlier access to flows of valuable information and extracting 'commission' for brokerage services by forging new ties and linking its unconnected alters. From the outset, Granovetter's argument for the strength of 'weak ties' and Burt's view of the value of 'social capital' for innovation is in accordance with Wellman's view of 'glocalised' network communities, and Saxenian's findings of the value of transnational communities and networks over great geographical distances for sustaining continuous innovation and competitiveness: for example, in Silicon Valley and Hsinchu global technology regions of ICT RTD (see Section 2.2). On the other hand, Coleman's view of social capital is clearly inward-looking compared to Burt's view. Coleman's argument about the value of 'cohesive' strong ties for learning and innovation resembles the argument put forward by Wenger in his conceptualisation of organisations as a constellation of tightly knit, overlapping and inward-looking communities of practice (see Section 2.3).

3.6 Summary

Section 3.1 introduced recent non-linear models of time-based innovation, highlighting the importance of collaboration networks as the social locus of knowledge-intensive technological innovation. A typology of innovation models based on Dodgson (2000) and Rothwell (1992) has identified five generations of models describing innovation processes. Of particular importance is the last generation of systems integration and networking models that has placed a premium on speed and flexibility for shortening lead times, and facilitating collaboration with lead customers and primary suppliers through horizontal linkages such as research consortia and joint ventures. Unlike early models that described technological innovation as a linear process of technology push, or market pull, or a coupling process of the R&D and marketing functions, the fifth-generation models have a non-linear nature as a result of their 'network' architecture, use of emerging Internet technologies and 'electronic toolkits' facilitating information flows across firms and other knowledge-generating organisations such as universities.

The importance of collaboration and networking for knowledge-intensive innovation was further illustrated by introducing the shift from 'mode 1' to 'mode 2' knowledge production (Gibbons *et al.* 1994), and the triple helix model of university–industry–government relations (Etzkowitz and Leydesdorff 2000). According to the former, the dynamics of science and technology in mode 2 reflects a socially distributed knowledge production system that has emerged alongside the traditional disciplinary structures of basic research and applied technology embodied in mode 1. The cross-disciplinary and heterogeneous nature of mode 2 knowledge production is also illustrated within the triple helix model focusing on the increasing interrelationships among different institutional spheres, i.e. university–industry–government. In particular the role of academia is worth highlighting in the mode 2 and triple helix models as universities and other knowledge-generating institutions cross over to industry and develop entrepreneurial and symbiotic relationships with private firms and government agencies alike.

The role of government in facilitating collaboration and networking among firms and other organisations was explored in Section 3.2 where the focus was on government innovation policy in Europe. The role of the EC, in particular, has been crucial for the past two decades (1983–2003) in fostering collaboration networks among firms and with universities and other research institutes with respect to ICT RTD work. The Esprit and IST programmes have supported more than 2000 research projects throughout Europe for carrying out collaborative research in ICT and paving the way for the emergence of the information societies and knowledge-based economies of the early 2000s. However, as was discussed above, the EC kept the emphasis of ICT research on pre-competitive 'blue-sky' research in the 1980s, and mainly supported national champions in the run-up to the single European market in 1992. During the third and fourth Framework programmes in the 1990s, however, it abandoned the notion of pre-competitive research in favour of collaboration in competition and participation of universities and other knowledge-generating institutions interested not only in hard science, for example computer science, but also socio-economic research (Assimakopoulos *et al.* 2000). Since the late 1990s, when the IST programme was launched, there has been an even greater shift towards involvement of small firms and users throughout the ICT value chain (see Table 3.1).

Since the early 1980s, in parallel with government policy makers, private sector managers have increasingly realised that to create and sustain competitive advantage in high-velocity markets such as ICT they have to acquire timely necessary knowledge and external capabilities from other rival firms and non-rival knowledge-generating institutions such as universities. The significance of distributed technological and dynamic capabilities was therefore introduced in Section 3.3 in conjunction with new forms of organisation and a relational view of strategy. Coombs and Metcalfe (1998) highlighted the importance of distributed technological capabilities within a network of collaborating and/or competing firms as the key for time-based innovation strategies throughout the late 1990s and early 2000s. Moreover, Eisenhardt and Martin (2000) have discussed the significance of dynamic capabilities such as product development and strategic decision making with regard to alliances in high-velocity markets (see Table 3.2). From the outset a relational view of strategy has shifted the emphasis of the discussion from individual firms to the inter-organisational networks that much of a firm's critical capabilities may stem from and depend on.

Technological innovation increasingly depends on key individuals and their personal networks as much as inter-organisational networks in high-velocity markets. Section 3.4 therefore discussed personal networks of a social nature fostering time-based innovation at the individual rather than the firm or community levels of analysis. Recent research has highlighted the significance of social networks in bringing information useful to RTD work and innovation across the organisational boundaries (Soh and Roberts 2003). Since the early 1990s, Freeman (1991) has introduced the term 'networks of innovators' for illustrating the vital importance of boundary-spanning contacts and other external collaboration networks for developing new products and processes.

The problem, however, with much of this research in the economics of innovation is that the methodology for classifying and measuring networks of innovators has been developed outside the scientific community linked to the economics of innovation invisible college. Sociologists such as Granovetter (1973, 1982), Burt (1987, 1992, 2005) and Wasserman and Faust (1994), rather than economists, have developed the SNA approach including concepts and methods that are used in the next chapter for analysing personal networks and their role in ICT RTD-related innovation throughout Europe and worldwide. In Section 3.5, therefore a number of the key arguments from SNA were reviewed, including the value of personal networks in building strategic alliances, the strength of weak ties in exchanging useful new information for innovation, and the value of structural holes and social capital for brokering opportunities and sustaining competitive advantage.

4 Ptolemaic views of personal networks in cross-national innovation

4.1 Introduction

This chapter presents the empirical findings and analysis of 10 case studies carried out in a comparative framework from 1998 to 2000. Each one of these RTD projects was supported by the EC, during its third (1990–4) or fourth (1994–8) Framework programme, bringing together a formal network of project partner organisations throughout Europe. The Appendix (p. 200) explains the underlying quantitative and qualitative methodology and presents a set of measures used in the analysis of findings and comparative evaluation in the subsequent sections of the chapter. Sections 4.2 to 4.11 present the main findings focusing on key ICT areas such as the design of new microchip architectures for computers, mobile phones and other hardware electronic devices; software application areas ranging from manufacturing to banking (i.e. security in e-commerce), utilities (i.e. electricity, gas), aerospace and defence; and projects addressing softer issues, such as the dissemination of research results and consensus building for trading intellectual property rights on the web worldwide.

The main assumption here is that despite a Euro-centric RTD policy from the EC trying to build a 'fortress Europe' the network ties which bind European collaboration in ICT research and make it work are to some extent exogenous to the collaboration, are personal and informal, and connect European partners not directly, but via unanticipated contacts in the United States and worldwide (see also Sections 2.2, 2.3, 3.2, 3.3). For each case study, three-dimensional computer-animated maps are put together describing the socio-technical configurations of the project network and analysing both its formal and informal structures. Specialised SNA and visualisation software (Borgatti, Everett and Freeman 2002; Richardson and Presley 2001) is used to produce these maps at the inter-personal level of analysis. Section 4.12 evaluates in a comparative framework the importance of informal networks for knowledge creation and exchange in such 'distributed' and dispersed networked innovations. Moreover, it discusses these cases from national and global perspectives with respect to ICT innovation policies linked to collaboration and competition throughout Europe and worldwide. Finally, Section 4.13 is the chapter summary.

4.2 Low-power asynchronous system chip for RISC Machines (OMI/DE-ARM Amulet2)

Amulet2 was a three-year Esprit III project carried out from 1992 to 1995 within the Open Microprocessor Initiative (OMI) / Deeply Embedded (DE) applications area. It produced a prototype demonstrating the feasibility of asynchronous embedded core microprocessors based on the ARM architecture. Advanced RISC Machines (ARM, www.arm.com) based in Cambridge, England, was the main contractor for Amulet2. Today ARM is a world leader in microprocessor cores and peripherals for computing, communications and consumer electronics applications. ARM's products are used in a broad range of industries from semiconductors to wireless telecoms, imaging and video games, palmtop organisers, smart cards, and automotive and computer networking equipment.

> ARM is the industry's leading provider of 16/32-bit embedded RISC microprocessor solutions. The company licenses its high-performance, low-cost, power-efficient RISC processors, peripherals, and system-on-chip designs to leading international electronics companies. ARM provides its partners with a total technology solution comprising cores, tools, platforms, and intellectual property components required in developing a complete system.
>
> (www.arm.com, accessed 23 January 2003)

ARM spun out of Acorn and Apple Computers' collaboration efforts in 1990 with a charter to create a new microprocessor standard. From a start-up company with a 12-person research team in 1991, the company grew rapidly to employ 160 people in 1996, and 350 people in 1999, with 20 sites in the United Kingdom, the United States, France, Belgium, Germany, Israel, Taiwan, China, Korea and Japan in 2003. The company has been involved in 14 Esprit projects, of which seven were carried out in Esprit IV. ARM was the main contractor and coordinator in about half of them. Typically, ARM was involved in an Esprit collaboration as technology provider, together with a software house, an equipment manufacturer and an end user to exert the necessary market pull. Esprit helped ARM in the early 1990s with vital seed funding as well as with developing links with original equipment manufacturers, end users and universities.

With the Department of Computer Science at the University of Manchester, ARM created the world's first asynchronous computer chip – Amulet – in 1994. Work on the asynchronous ARM architecture did not stop there. In 1996 the Amulet team delivered the Amulet2 embedded system chip prototype based on the Amulet2 project. Unlike the majority of current electronic products which use synchronous technology, the Amulet2 core is based on asynchronous technology that allows components to operate at their own speed and pass on results when they are ready, resulting in significant power savings as idle components can switch themselves off until required. Asynchronous design also demonstrates improved electromagnetic compatibility since asynchronous operation spreads current consumption over both time and frequency domains.

> Because things are changing so fast, you will pick up the best contacts you have from a job in the next project, maybe the devil you know ... You will go back, look at the guys who worked well ... you get the new guys that you need ... so I don't think that you would carry forward the same network through this fast changing process. I think you would review it and I think that is what we are describing we do. We are reviewing the situation – who we wish to work with and it is more likely to contain people we have worked with in the past and know well than a new one because the risk is lower.
>
> (ARM manager)

It is instructive to analyse a Ptolemaic view of the Amulet project network. Personal network data were collected from a professor of computer science leading the Amulet project at the University of Manchester (see Graph 4.1), who identified those who had supplied him with information of importance to the project: individuals within his own team at the University of Manchester and from ARM, and two engineers from Hagenuk Telekom in Germany. In subsequent rounds of nominations all these individuals reciprocated the personal tie to the professor in Manchester. Note that the spheres represent individuals in Graph 4.1 and the ties represent nomination network data. The size of the spheres varies according to degree, closeness and flow betweenness centrality (see Appendix) and the tics only connect

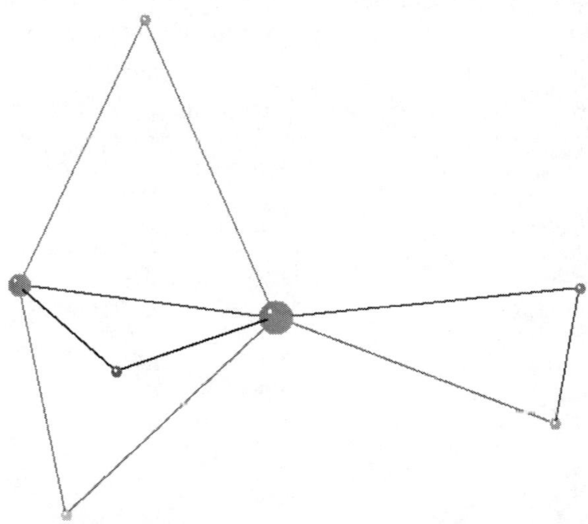

Graph 4.1 Amulet2 personal networks.

people within the formal EC-funded project network. The most central individual in this network is the professor at Manchester (in the middle of Graph 4.1). However the second most central individual is not an ARM man but a German engineer from Hagenuk Telecom (middle left-hand side of Graph 4.1).

> The value of networking is that you don't build a network of people that don't know what you are doing. X is a product that came out of a very significant piece of work done by a side group in a European project about things in a completely uncontrolled way ... Yes, it is a spin off ... spin off which you cannot predict and has actually been very identifiable and beneficial and it is a bit strange ... we were not working on that ... we were working on a different aspect of the problem ... You could argue that it was a result of the first problem that you thought of, and you could equally argue it was not ... so EF brought in Hagenuk, at the end because this other company (TKI) fell out. Hagenuk met ARM and started to use ARM a lot. In the project was TG [a university professor at Manchester University] on an asynchronous version of the ARM design. Hagenuk started thinking about that. Suddenly they discovered that the Manchester design performed well and you have collaboration between Manchester and Hagenuk that definitely was not intended at the start of the project.
>
> (ARM manager)

Overall the pattern of relational data of TG's personal network shows four cohesive subgroups, or cliques, at Manchester, Hagenuk Telekom and ARM with no connection to other organisations within the formal Esprit collaboration. The entire Amulet2 project network is shown in Table 4.1. Twelve companies and universities participate in this formal project network.

Table 4.1 Amulet2 project network – partner organisations by country

Partner organisations	*Country*
ARM Ltd (Main partner/contractor)	UK
Eletronica	Italy
Etnoteam	Italy
GEC Plessey	UK
Hagenuk Telecom	Germany
IRIS	Italy
OPSIS	France
Philips Kommunikations	Germany
UMIST	UK
University of Manchester	UK
University of Hannover	Germany
VLSI – Technology Centre de Recherce Europeene	France

The professor at Manchester University, however, has personal connections with only a quarter of the partner organisations in Amulet2. He has no important information links with partners, for example in France and Italy, and no links with many UK partners, including academics at UMIST (University of Manchester Institute of Science and Technology) next door. The link between him and ARM is personal, broad and dates back to Acorn in the early 1980s, long before ARM was set up in 1990.

Similarly, ARM managers highlight that the company's formal collaboration with its partners takes time to develop, something which the EC does not always seem to appreciate.

> This German company [Hagenuk Telekom] approached us to say they were interested in using the technology in telecom controls. So they came to us ... We worked together in a previous [Esprit] project ... which gave them some confidence in us, which meant that they approached us for the next project [Amulet] where we actually worked together. I think there is a very important lesson in that: Esprit trying to enforce collaboration in projects and trying to get everybody working on saying yes ... What has worked for us is to be in a project with potential collaborators but working on different things, and watching each other, and then, when we recognise there was some synergy because we had been working in the same project we had the mutual confidence and trust that we work together in the next project. If we had been required to be collaborating and working on the same thing in the first project we would probably never have got together. They would not have known us. We would not have known them. Before you marry somebody you have a period of engagement, you meet, you go to parties together. In a sense, Esprit has sometimes felt like it was trying to force people into marriages before they actually got to know each other.
>
> (ARM manager)

So partners need time to get to know each other before they commit themselves to a formal collaboration such as the one presented above.

> Our particular contribution to Amulet is carried out in the context of a global activity and we do have contacts with individuals all around the world in that area. Who are the key individuals in that area? ... There are a few hundred people. The community is highly integrated. I think of it in terms of groups of individuals. Here we have twenty people in my group, and there are half a dozen people in Holland. In this country there are three or four people at Southbank, four or five people in Newcastle, a group of four or five people in Cambridge and a number of other smaller groups throughout the country. Then if you go outside this country ...
>
> (TG, University of Manchester)

It is therefore clear that the Amulet2 project team is part of an international technological community of two to three hundred people working within the tradition of practice of asynchronous logic and low-power problems.

> We were actually talking to a much wider community than our project ... What I would do next time round, hope to do next time round, [is] have [a] smaller core project, but we will have a formal structure to involve outsiders ... there is a core partnership doing the work, but we have invented an affiliate structure and those people will be involved in the evaluation conferences, those people will be involved in the assessment of the development work, and eventually will be asked to put their hands to a memorandum of understanding.
>
> (ARM manager)

4.3 Development of Libraries and Physical Models for an Integrated Design Environment (Delphi)

Delphi was a three-year Esprit III project carried out from 1993 to 1996. It aimed to address issues related to the accurate prediction of the temperature of critical electronic parts at the component, board and system levels; and also to create and validate 'detailed' 3-D conduction models for thermal analysis at all packaging levels. Component thermal management is becoming more and more crucial as ever more transistors are incorporated into single pieces of silicon and applications require more computing power and ever faster processors. The main contractor for Delphi was Flomerics (www.flomerics.com), a start-up company founded with venture capital in the late 1980s. Today, Flomerics is a public company and world leader for electronics thermal analysis in a broad range of industries from semiconductors, to computing, telecoms, consumer electronics, aerospace and defence, based in Surrey, in the south-east of England.

> Flomerics has a very strong product (Flotherm) for process-orientated design issues. Companies, such as Intel and HP, who depend on Flotherm have already reduced short design cycles by several weeks. Flomerics has been retained by these companies because it is the industry benchmark standard for electronic thermal models. This means that there is a huge community using and improving the software, building a vast library of standard models that can be dispersed across departments and suppliers.
>
> (www.flomerics.com, accessed 23 January 2003)

The Delphi project network also included the following formal partners: Alcatel-Bell (Belgium), Alcatel-Espace (France), National Microelectronics Research Centre (NMRC) at the University College Cork (Ireland), Philips-CFT (Netherlands) and Thomson-CSF (France). Graph 4.2 shows the personal network of the Delphi main contractor/project manager intertwined with the personal networks of other key players for the project. Internal ties connect people within the formal EC-funded project network and are dark grey, while external ties connect people from outside the official project network and are light grey.

The first striking finding is that overall more than half (56 per cent) of all people shown in Graph 4.2 and almost two-thirds (62 per cent) of all nominated ties are external to the formal project network. Moreover, the most central individual in the network (regardless of how centrality is measured) is a Dutch electronics engineer in Philips Research Labs at Eindhoven. Second most central is a Belgian engineer at the Alcatel-Bell research division on thermal compatibility in Antwerp, and third is the project manager from Flomerics, the UK engineer who nominated the Dutch and Belgian engineers in the first place. What is even more interesting is that more than half of the Dutch engineer's personal network is outside the Esprit formal agreement. His network includes sources of information essential to this Esprit project in the United States (for example, at Stanford University, the University of Minnesota and Motorola). In Europe, outside the project, crucial information came from sources in Philips Semiconductors in Nijmegen, Siemens Semiconductors in Munich, and SGS-Thomson (currently ST Microelectronics in Milan).

It is also notable how nominated sources outside the Esprit project themselves nominate sources of information within the project so that networks which might have been thought to have been internal to Esprit are in fact intertwined with external information networks. The extent of overlap can

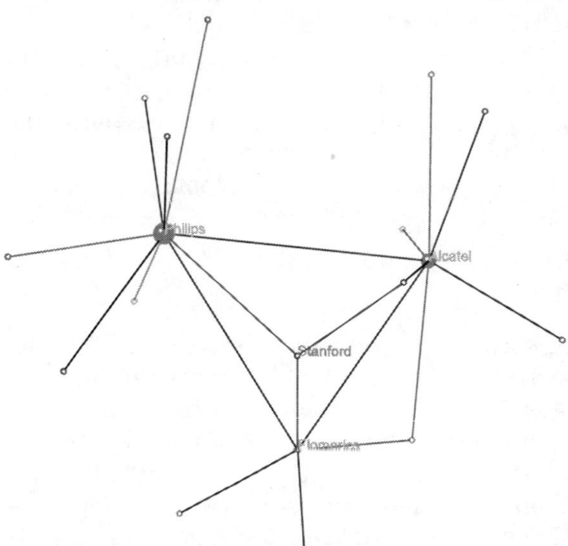

Graph 4.2 Delphi personal networks.

be seen in the case of an American professor (bottom middle of Graph 4.2) from Stanford University who is linked with the main UK contractor, but also with two other nominations of the latter: the Dutch and Belgian engineers mentioned above. Such overlaps allow valuable information for Esprit innovation to flow back and forth across the Atlantic via a number of direct and indirect ties within and outside the project consortium.

Besides centrality, the structural holes measures of everyone involved in the Delphi network were also calculated with Ucinet (Burt 1992). As expected, the effective size of the Dutch engineer's network had higher scores than the Belgian and UK project managers. More detailed analysis of the cohesiveness of Graph 4.2, however, reveals that there are only three cliques (triads of mutually nominated engineers) from Flomerics, Philips and Alcatel; Flomerics, Philips and Stanford University; and Flomerics, Alcatel and NMRC at University College Cork in Ireland. This finding highlights the role of the UK project manager as he is the only node of the project network who participates in all three of its cohesive subgroups. The UK project manager knew his colleagues at Philips and Alcatel as customers of Flomerics for about five years before the Delphi project. In fact the initial idea for the project mainly came from these 'lead' customers: end users who later added value to the project and did not pose any competitive threat from its inception.

> Yes the idea came to us from our customer base. Philips is consumer electronics but its part involved in Delphi was a sort of central research group, not quite technology development, Alcatel Bell is a telecoms company, satellites in space, Thomson into professional electronics, radar defense, they were quite big customers of ours and have been for a number of years in fact prior to the Delphi project ... fairly closed knit group. We do share very much our software in common ... Yes correct NMRC is a research organisation. No competition ... which is one of the reasons it was quite successful.
>
> (Flomerics manager)

4.4 In-line Monitor for Process Optimisation and Verification (IMPROVE)

Improve was an 18-month Esprit IV project from 1996 to 1997. It was a semiconductor equipment assessment project aimed at assessing the performance of an instrument produced by SOPRA (www.sopra-sa.com), a French world leader in the photonics industry, for the United Kingdom's Defence Evaluation and Research Agency (DERA). The SOPRA Multi-Layer Monitor (MLM) is an optical monitoring instrument based on multi-channel spectroscopic ellipsometry (SE) used in the semiconductor industry for product design and manufacturing wafers with a diameter from 100 mm to 300 mm. According to SOPRA's website:

> The strong advantages of ellipsometry are its non destructive character, its high sensitivity due to the measurement of the phase of the reflected light, its large measurement range (from fractions of single layers to micrometers), and the possibilities to control in real time complex processes. We must distinguish between single wavelength ellipsometry which can measure only two parameters and SE which can analyse complex structures such as multi-layers, interface roughness, inhomogeneous layers, anisotropic layers and much more. We are interested here only in SE.
>
> The MLM's ability to map very thin multi-layers on product wafers provides an advanced measurement capability for state of the art processing. This removes the need for expensive monitor wafers which are often non-representative. Optical diagnostic methods are the foundation of semiconductor metrology. The most common tools are optical microscopes and particle counters. SE should be used as the next line of analysis due to its flexibility and non-destructive application. The SOPRA MLM Spectroscopic Ellipsometer has an optical multi-channel analyser (OMA) and a micro-spot capability. This enables it to be used for measurements on very thin composite layers on product wafers. It can be used in analytic (scan) mode for accurate calibrations whilst the OMA enables it to measure 4 parameters at 9 sites on three layers with high throughput, typically 35 wafers per hour.
>
> (DERA project leader)

According to *Prosoma*, SE can measure multiple buried layers because of the wavelength dependence of the optical penetration depth. The MLM is capable of measuring, non-destructively, up to five layers simultaneously on patterned product wafers, providing information on numerous parameters such as thickness, composition and roughness. The major result of the project was to demonstrate the ability of the SOPRA MLM to detect 'in-line' process-induced variations reliably, rapidly and non-destructively in complex multilayer structures. MLM data correlated extremely well with the more time-consuming data obtained from X-ray diffraction, transmission electron microscopy, etc. as well as with electrical data from processed devices.

> DERA has many materials and device programmes which require non-destructive in-line monitoring of the cross-wafer uniformity of multilayer structures. Examples are advanced bipolar circuits and microstructures requiring microspot measurements on processed wafers. This project provides a unique opportunity to form relationships with IC manufacturers, to understand their requirements and collectively add value for competitive manufacturing. Our extensive range of optical and complementary characterisation facilities makes DERA a true 'assessment' site with the ability to provide corroborative data to assess accuracy.
>
> (DERA manager)

This work forms part of DERA's broader activities aimed at developing in-line, in situ and real-time process control using optical techniques and advanced analysis algorithms, in particular for application in commercial systems as a demonstration of the 'minifab' concept. The main business areas where the results of the project are shown to have an advantageous impact are product development and improvement; quality control and conformance monitoring; statistical process control for yield enhancement; and attainment of new data relating to product in a non-destructive rapid manner. The outcomes of these advantages are reduced product development times, increased productivity and yield, reduction of monitor wafer usage, and measurements on actual product resulting in an overall reduction of product cost and a competitive edge in both manufacturing and product development.

> SE can be used for many problem-solving activities before going on to more complex, time consuming and destructive methods. Our understanding of this approach means that significant benefits can be gained in the manufacturing environment. Having a tool able to give results in minutes and hours rather than days is very advantageous. SOPRA sees the advantages of DERA as being able to test the MLM for many kinds of multilayer samples, to develop an analytic database and have the authorisation to disclose the results. Users and equipment manufacturers can work openly for mutual benefit and new ideas can be implemented to improve the equipment. Many systems have already been purchased and this is a unique opportunity to test more advanced solutions and ideas with less commercial pressure, and to understand requirements for the next generation technology. A good, friendly cooperation with users has been established.
>
> (DERA manager)

Besides DERA and SOPRA, the formal project network also included three semiconductor companies: GEC Plessey in the United Kingdom (now Mitecsemi); TEMIC in Germany (now ATMEL); and AMS in Austria. Graph 4.3 shows the personal network of the Improve main contractor/project manager at DERA intertwined with personal networks of other key players for this project. As before, the size of spheres varies according to centrality measures and the same colour of ties reflects the fact that all ties are internal to the formal project network.

The first striking finding from Graph 4.3 is that the DERA engineer/project manager is the third most central individual in the network. The technical leader at SOPRA (at the bottom of the graph), who was nominated by the DERA engineer in the first place, is the most central individual of the network, followed, surprisingly, in second place by the EC project officer (top middle of the graph) who was nominated by the SOPRA manager. It is worth noting how the SOPRA engineer highlighted the significance of informal face-to-face contacts:

> The new technologies, even if they are faster than traditional means of communication, will not replace a direct face to face meeting. In France, lunch or dinner have always been important to judge if someone is trustworthy … if we can trust, if we like and value this relation, and of course the reliability of information exchanged. At least 45 min are necessary to chat before any useful information has been exchanged from my experience. E-mail contacts cannot do this.
>
> (SOPRA manager)

Another key finding of Graph 4.3 is that after analysing cohesiveness eight cliques/triads of nominations are identified. The EC project officer participates in most cliques – six out of the eight – the DERA project manager participates in five out of the eight cliques, and the SOPRA project manager participates in half of these cliques. The relatively high number of cliques in the graph provide several overlapping paths for information exchange, not only with the EC project officer but also among the key players in DERA, SOPRA and the semiconductor manufacturing companies, namely TEMIC and AMS. Moreover, several other contacts are nominated within the formal project network, including individuals who work throughout Europe in relevant organisations, such as the semiconductor equipment assessment group at the Rutherford Appleton Laboratory near Oxford, the Fraunhofer Institute at Erlangen, and the Joint Equipment Manufacturers Initiative in France (www.jemi-france.org) a non-profit association of manufacturers

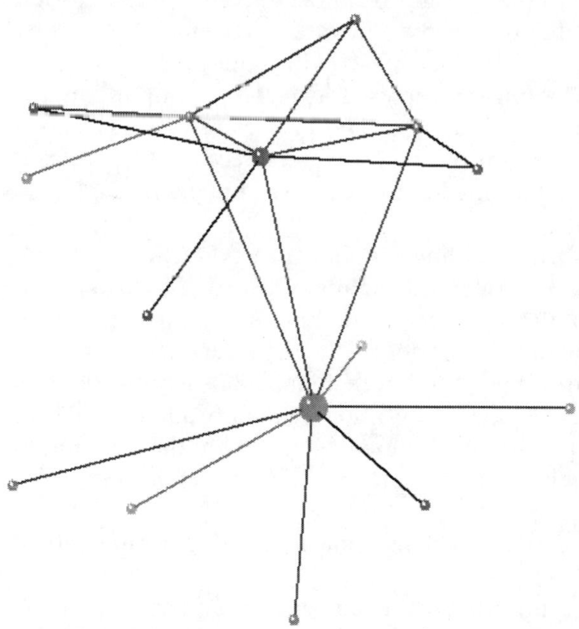

Graph 4.3 Improve personal networks.

and suppliers of equipment, materials and services to the worldwide semiconductor industry. All these additional contacts, however, are isolates in Graph 4.3.

> They behave primarily as individuals. They are there to champion their company's interests, but what they can do and what they are able to do within the project, and on what timescales they are able to do it, is much more down to them as individuals than it is the company.
>
> (SOPRA manager)

4.5 Feature-based Integrated Rapid Engineering System (FIRES)

Fires was a three-year Esprit III project carried out from 1992 to 1995. It aimed to develop and validate a software prototype of a computer-aided design/manufacturing (CAD/CAM) product modeller focusing on the requirements of three-dimensional (3D) complex parts in manufacturing. According to *Prosoma*, it prototyped a new generation of computer-aided engineering (CAE) tools which supported an integrated approach to the rapid modelling, analysis and manufacturing of complex mechanical engineering components. The integration of these three disciplines – modelling, analysis and manufacturing – has enabled engineers to optimise designs and keep manufacturing costs to a minimum for a broad range of industries from machining and tool making to the automotive industry, aerospace, plastics, footwear and packaging. Validation of the software has been carried out by demonstrating the results on 3D milled parts with high-precision manufacturing requirements and using examples drawn from components for the tool design and automotive industries. The project results have been incorporated in software marketed by the Fires' main contractor, Delcam International (www.delcam.com) headquartered at Birmingham, England.

> Delcam is one of the world's leading suppliers of CAD/CAM software for product development throughout the manufacturing industry. It is the largest developer of product design and manufacturing software in the United Kingdom, with subsidiaries in North America, Europe and Asia. Delcam's software is used by more than 9,000 organisations in over 80 countries to boost productivity, improve quality and reduce lead times, and so increase profitability at every stage, from the creation of a concept design through to the manufacture and inspection of prototypes, tooling and sample components.
>
> (www.delcam.com, accessed 23 January 2003)

Delcam International was established at its current headquarters in 1991, though the origins of the company can be traced back to Cambridge in the late 1960s. Donald Welbourn, director of Industrial Cooperation at

Cambridge University, saw the possibility of using computers to assist pattern makers to solve the problems of modelling difficult 3D shapes. 3D modelling of complex parts is taken for granted today, but only crude 2D drawing systems were available back in the late 1960s and early 1970s, using terminals linked to large mainframe computers. Initial work at Cambridge was sponsored by Ford in the United Kingdom and Control Data and Delta Engineering Group in Germany with back-up from its largest customers Volkswagen and Daimler Benz.

A decade later, in the early 1980s, the technology began to be competitive and the development continued at Cambridge together with a small team established at Delta Engineering in Birmingham in 1977. By 1984 the Birmingham team was larger than that at Cambridge. The unique features of its software, which was developed from the beginning with an equal emphasis on design and machining, made it different from its competitors and this was appreciated by users. Most other systems had developed 2D drafting first, followed by 3D modelling with machining added later. In 1989, the company was bought from Delta Group in a management and employee buy-out and was renamed Delcam. It was floated on the alternative investment market in 1997. Delcam participated in a number of Esprit projects in the 1990s, and earlier, in Esprit II for example, in Ipdes – the Integrated Product Design System project led by CETIM from 1989 to 1992.

The network of Fires' formal partners included the following organisations: Centre Technique des Industries Mécaniques (CETIM, France), CIMDATA (Germany), MARES and Associacion de Investigacion Tekniker (Spain), and the Technical University of Darmstadt (Germany). According to *Prosoma*, the Fires prototype enables users to do the following: design mechanical engineering components more quickly than with traditional CAD; perform analyses on these models quickly and automatically; feature definition, forces, temperatures, fixed points, etc.; generate a process plan from the CAD model quickly and automatically, showing the manufacturing processes required, and optimise according to specified criteria (e.g. time, cost); alter the process plan according to shop-floor criteria; automatically generate computer-numerically controlled data to manufacture the component from the process plan; and represent machining operations graphically on-screen. The three main models – the Feature-based Modeller, the Process Planner and the Analysis Interface – are developments of pre-existing software packages. Fires' innovation is in the integration of these components via neural data structures that use information models based on STEP Resource Models and Application Protocols.

> Yes, over the last year or so we have put quite a lot of effort in to making contacts in Brussels and a number of us here have acted as evaluators or reviewers for project proposals. The EC does pay us for it. But in reality the money is not very good and we would not do it for any other reason than just to make those contacts make it easier to get further funding.

> I don't know whether it will make it easier or not, but it helps to get more knowledge of how everything works and understand the process. For example, to do an evaluation on projects helps you to understand very well what the criteria are, and how projects are judged so you stand a much better chance when you write a proposal of getting it accepted. I don't think there is too much politics involved in choosing projects. Before we started off we imagined that most projects were chosen on political rather than technical grounds, but I don't think that is really true. In my experience of evaluating projects I was quite impressed how unbiased and how the project seemed to be evaluated purely on its technical merits. So I think the real advantages just come from understanding that process well. Understanding what people look for when they are judging projects and how to present a project clearly.
>
> (Fires manager)

Graph 4.4 shows the personal network of the Fires main contractor/project manager intertwined with the personal networks of other key players for the project. As before, the size of spheres varies according to centrality measures, and the colour of tie reflects whether a tie is internal (dark grey) or external (light grey) to the formal project network. The first striking finding is that the most central individual in the network is a Brazilian engineer at the University of São Paulo (upper right-hand corner of Graph 4.4) who carried out his doctoral degree at the Technical University of Darmstadt in Germany and was part of the earlier integration manufacturing Ipdes Esprit II project. The Delcam project manager, who nominated the Brazilian engineer in the first place, is the second most central engineer in

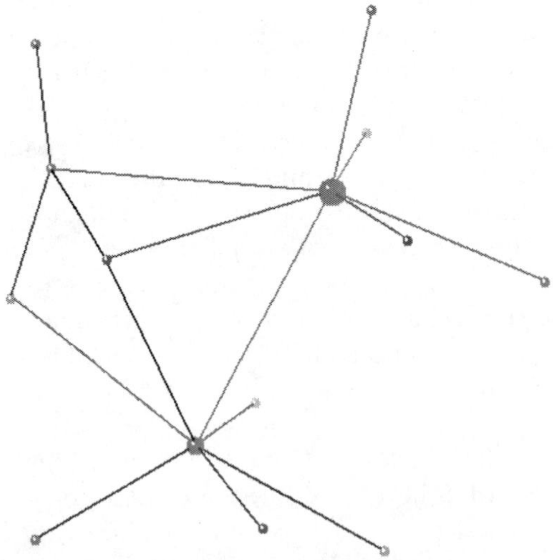

Graph 4.4 Fires personal networks.

Graph 4.4. There are also only two cohesive subgroups or cliques in Graph 4.4, connecting engineers in Delcam with the universities of São Paulo/Darmstadt and a French engineer in Kadetech near Lyon.

> Universities are always very keen to be involved in projects and we get contacted by universities a fair bit. Looking at projects to be involved in, because of our research in CAD, there are a number of universities who are working on, for example, surface modelling and surface mathematics and they tend to have links with us. Perhaps try to place research students here, or they place people at the end of their courses. Universities are not much of a problem ... End users are also quite easy because we have more than 9000 customers who are all end users, so it is pretty easy for us to find customers in any country of the world that will act as end user partners.
>
> Probably the more difficult partners to find are other developers. If we want to have two software companies in the project so that it appears to be pretty competitive then it will be difficult to find companies who want to share knowledge with ... quite often we are dealing in direct competition with that company and would not want to work together ... so the sorts of things that we tend to do is work with a company that is not a direct competitor, for example, in the area of data exchange, where working with a competitor would be very useful and not threatening in any way. Clearly it would be great if we were allowed to work with partners in the United States as part of these funded projects but the whole point is to treat the Americans as competitors and try to be more competitive and have a certain technological advance on Americans as far as these funding bodies go.
>
> (Fires manager)

It seems that the Fires project manager clearly identifies the benefits of working with competitors – other CAD/CAM software firms, in particular those located across the Atlantic – but the institutional barriers created by the EC and Esprit rules for allowing such participation, on top of the obvious market barriers, are formidable and prohibit such collaboration. End users, such as 'lead' customers (Von Hippel 2001) and universities laboratories, do not pose any competitive threat and therefore are considered as legitimate partners for enabling continuous innovation. After all, as the next quote indicates, globalisation of innovation through lead users has benefits not only for organisations in Japan or the United States, but more importantly nearer home in the United Kingdom, or in other European countries such as Spain. How the policy makers in Brussels are going to accommodate these benefits in government-funded programmes such as Esprit/IST is still a challenge that has to be addressed for innovative firms such as Delcam.

> Last week I was in Japan visiting some of our sales partners there and

our customers and it was very interesting to see how the mould makers in Japan differ from the mould makers in other areas, like Spain or England, and the different requirements that they have on the software and it is very useful to make those sort of contacts and to change our software in order to cope with the requirements of those sort of industries. Japan was particularly interesting because I think they are very much state of the art in terms of their work in mould making. It would be nice if we could do these sorts of projects with a Japanese partner. It would be great. But to my knowledge I do not think we can do that ... Still when you are producing a proposal you have to put a lot of effort into how this is going to give European advantage and as far as software vendors ... well it is not enough for us to say it gives us an advantage as software vendors, but we have got to say that it gives an advantage to our European customers ... so if we make a big thing in our proposals about how it helps our Japanese or our American customers, then it does not seem to be fulfilling that bit of the proposal very well. It is very political, you are right ... but on the other hand how you define who will benefit in Europe and who in the United States or Japan ... this is problematic. Inevitably because we are in Europe perhaps we have a bigger user base in Europe than in America or Japan but ... maybe it will help the European mould makers first, or quicker, but will inevitable help people in Japan and America too.

(Fires manager)

4.6 End to End Security over the Internet (E2S)

E2S was a 25-month Esprit IV project carried out from 1995 to 1997. It sought to develop enabling technologies for secure business-to-business transactions over the Internet. According to *Prosoma*, E2S technology is a major step towards ensuring the security of confidential information and commercial transactions over the Internet. The E2S architecture is based on secure electronic transaction (SET) technologies for bankcard payment systems. SET is an open standard developed jointly by Visa, Mastercard and their technology partners to enable card transactions to be made securely over open computer networks using encryption technology. In the mid-1990s, SET became available in Europe, thereby enabling European banks to take a leading role in the international development of secure electronic commerce for consumers. The main contractor for E2S was ANSA (www.ansa.co.uk) Architecture Projects Management (APM) in Cambridge, England.

The ANSA research project started in the United Kingdom a decade earlier than E2S with a focus on open distributed system architectures, network security and training of industry sponsors based not only in the United Kingdom but also in the United States and France. The ANSA architecture built on state-of-the art technology and open standards to enable application components to work together despite diversity of programming languages, operating systems, computer hardware, communication protocols and security policies, to name but a few of the serious issues hindering the integration of systems. ANSA was initially funded by the Alvey programme of the United Kingdom Department of Trade and Industry from 1985 to 1988 and included as sponsors Bellcore and Bell Northern Research Europe, British and France Telecoms, DEC, HP, GEC, ICL, etc. For its second phase of development ANSA was supported by Esprit II and its Integrated Systems Architecture project from 1988 to 1993. From 1994 to 1999, the ANSA project ran with full industrial funding, while APM carried out several Esprit III and IV projects in parallel, including E2S.

As well as APM Ltd, the formal E2S project network included as partners HP European Laboratories near Oxford (United Kingdom), HP Grenoble Networks Division (France), GMD National Centre for Information Technology (Germany), GemPlus (France), the Technical University of Berlin (Germany), VISA International (France) and, as associate partners, HP Worldwide Customer Support Organisation (WCSO), Smart Card Forum (United States), Onyx Ltd, and Swiss Bank Warburg (see Figure 4.1, taken from www.e2s.com, 15 December 1998).

Graph 4.5 shows the personal network of the E2S main contractor/project manager in APM with the personal networks of other key players for the project. The most striking finding is that the biggest sphere in the network (upper right-hand corner of Graph 4.5) is not the APM main contractor but the vice-chairman for electronic commerce in the Bank of America situated in downtown San Francisco, California. The second most central individual (middle of Graph 4.5) is an engineer at Bellcore Labs in New Jersey. It is also interesting that there is a contact at the Citibank Group in New York (bottom right-hand side, in the middle of Graph 4.5) who is common to both

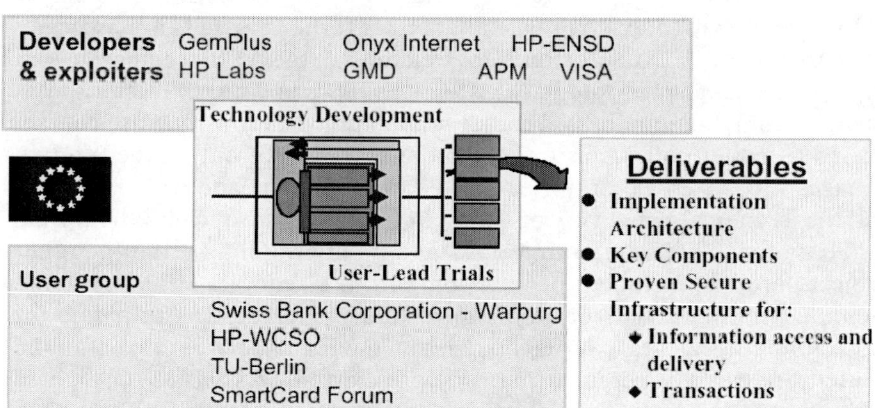

Figure 4.1 E2S project network.

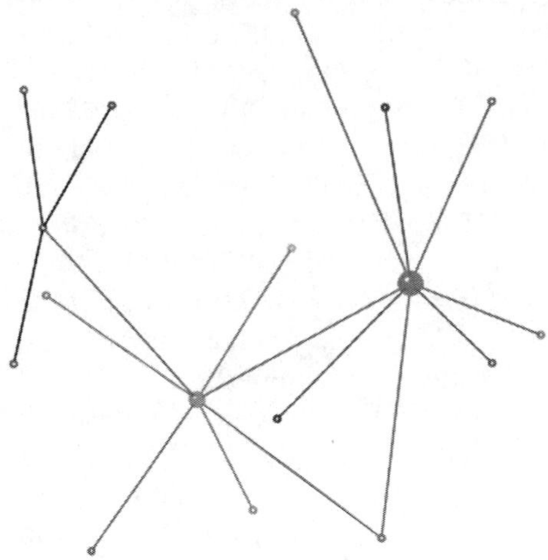

Graph 4.5 E2S personal networks.

the source in the Bank of America and that at Bellcore Labs. It would seem from the structure of this graph that the United Kingdom main contractor benefited most from personal contacts with individuals in the most dynamic parts of the IT world and, unsurprisingly for e-commerce in the mid- to late 1990s, these were located outside the European Union, primarily in the United States.

It is worth also noting that the only cohesive subgroup, or clique, in the E2S informal network is all together outside the formal project network, connecting US colleagues in the Bank of America in San Francisco, Bellcore Labs in New Jersey and Citibank in New York. Furthermore Graph 4.5 reveals that more than two-thirds of the personal network of the main contractor's technological leader lies outside the European Union boundary connecting him to individuals who are neither formal partners nor associate partners with the E2S consortium. The relations of the APM technical leader with his US colleagues date back to the late 1980s and are characterised as friendly, unlike the majority of other nomination relations that are characterised as either colleagues or acquaintances. Overall only three internal linkages – to the HP laboratories near Oxford, VISA headquarters in Paris, and the Technical University of Berlin in Germany – are within the formal collaboration. There are no internal linkages in Graph 4.5 from the other project partners, including people from HP in Grenoble, HP-WCSO and GemPlus in France. Therefore the most valued sources of information for innovation for E2S are external linkages on the East and West Coasts of the United States. 'My personal network pre-existed E2S work, but gained additional nodes through collaboration with new partners, or new groups in existing partners ... the nodes that are most important and active are the ones that developed into business relationships' (APM manager).

4.7 Time Constrained Integrated Management of Large Scale Systems (TIMELY)

Timely was a three-year Esprit III project from 1993 to 1996. It aimed to further develop and consolidate an innovative approach to the integrated time-critical management of large-scale energy transport networks within the application domains of electricity and gas distribution. At the heart of the project was the development of software for the timely diagnosis, prevention of failures and downtime resulting in accidents, loss of supply, environmental impacts, etc., in such critical infrastructures as electricity transmission over large geographical areas, e.g. northern Italy. According to *Prosoma*, effective management decisions within the energy management arena relate to three critical tasks which should be undertaken within severe time constraints and depend on: the detection of abnormal situations; the identification of their causes; and the determination of remedial actions. At the time these tasks were treated separately using different techniques ranging from human controllers to rule-based diagnostic systems based on expert decision support systems developed in the 1980s.

The project main contractor was Syseca (www.syseca.ch) a company based in Manchester at the time, and currently headquartered in Zug, Switzerland, which provided integrated software solutions within the energy arena.

> We have extensive experience in 'Real Time' systems development having designed and implemented world-class network control (SCADA) solutions for the electricity industry. Our highly tuned analysis, design and software engineering skills give us the ability to provide successful solutions in very challenging business environments. We tailor our software to meet the specific needs of our customers, though to avoid over-investment we utilise a mixture of existing Syseca software components, third party products and customer specific new development.
>
> (www.syseca.ch, accessed 17 December 2004)

> Timely was preceded by the Artist project in Esprit II. Partners in the Artist project approached Syseca for putting together the Timely project proposal in Esprit III. In fact both projects shared a number of partners in Italy, Spain and the United Kingdom.
>
> (Syseca project leader)

The Advanced Reasoning Tool for Model-Based Diagnosis of Industrial Systems (ARTIST) project was a 30 month Esprit II project from 1990 to 1993. Like Timely, it had aims and focus on the electricity transmission industry. It was based on a novel approach to diagnosis and early detection of developing faults called model-based reasoning (MBR), which was regarded as an evolutionary path to expert systems in the early 1990s. The

Timely project used MBR techniques to explicitly model distribution networks, thus building on architecture and tools for model-based diagnosis developed within the Artist project, but more importantly it extended these models to include detection and remedial tasks within an integrated MBR framework. The time-critical nature of these applications and tasks required that a mechanism for time-constrained reasoning for MBR had to be developed. Further multiple model descriptions, at various levels of abstraction and approximation, were used to provide a progressive reasoning capability that allowed the best possible diagnosis, and consequent remedy, to be produced within the prescribed time constraint of an application.

One of the products developed by Timely was called PowerDA and it is marketed by Syseca as an integral part of their electricity supply packages. PowerDA is a fault diagnosis and condition monitoring system for the electricity industry, helping utilities to manage their high-voltage distribution networks, for example during adverse weather conditions, ensuring the security of power networks and avoiding total power loss, which can take hours to bring back while leaving large numbers of consumers without electricity supplies. PowerDA has been validated against data from electricity utilities in Italy, Greece and Norway. According to Cordis, such software was a significant advancement and consolidated technical advances made in many previous Esprit projects in the arena of software intensive systems for large industrial applications.

Apart from Syseca, other formal partners in Timely were: Landis and Gyr, an energy management company at Zug, Switzerland; ENEL, the Italian electricity and gas utility; CEPSA, a large petroleum company based in Spain and around the world; and the Department of Electrical, Electronic and Computer Engineering at Heriot-Watt University in Scotland. Graph 4.6 shows the personal network of the project leader at Syseca together with

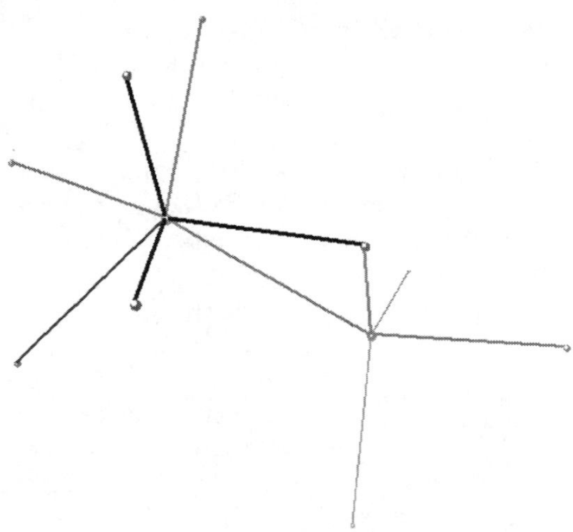

Graph 4.6 Timely personal networks.

a few other personal networks of people who contributed important information for innovation in Timely. The same conventions apply as above for representing individuals and ties showing nomination network data.

The most central individual in the network is the main UK contractors themselves: the largest sphere at the upper left-hand side of Graph 4.6. However, what is even more interesting is that part of their network includes sources of information essential to this project within the consortium, such as the contact in ENEL (sphere in the upper right-hand of Graph 4.6), and also outside the formal project network in other user organisations, such as the Greek Public Electrical Company (lower right-hand side of Graph 4.6). In fact, the only clique that exists in the graph connects the project leader at Syseca with his colleague engineers at the lead electricity distribution organisations in Italy – ENEL – and Greece – Public Electrical Company.

> It is my belief that it is of importance at the present time, with the evolution of digital technology in practically all fields of the electricity supply industry, and in particular for security and control engineers involved in the utility side, to network and share information and knowledge through international projects and institutions such as IEE, as well as through personal informal networks.
>
> (ENEL manager)

4.8 Reducing the risk of gas explosion; preparing the FLACS code for commercialisation (Flacscom)

Gas explosion is a major safety risk on petrochemical production and processing facilities both on and offshore. Flacscom was a 14-month Esprit IV project carried out from 1996 to 1997. It aimed to disseminate and commercialise the Flame Acceleration Simulator (FLACS), a sophisticated software code with computational fluid dynamics (CFD) as its main basis, for reducing risks related to gas explosions in the petrochemical industry. According to *Prosoma*, CFD codes have been shown to be superior to simpler methods, such as venting guidelines and models for predicting the consequences of explosions. FLACS is by far the most robust of the CFD codes, having been developed through a collaborative effort between companies within the offshore oil and gas industry. Note that FLACS is the result of almost two decades of intense fundamental research, with a total expenditure exceeding €15 million up to 1996 (and the research is still ongoing) into gas explosion prevention primarily on offshore platforms.

> FLACS calculates the explosion pressure and other flow parameters as a function of time and space for different geometries and explosion scenarios. It takes account of the interaction between flame, vent areas and obstacles such as equipment and pipe work. The full gas dynamic partial differential equations are solved for a set of control volumes,

> including the effects of turbulence and chemical reactions. The build-up of flammable gas clouds, the impact of improved ventilation on the cloud build-up and the gas explosion process and its consequences can all be investigated. The effect of mitigation techniques, such as water spray, can also be modelled as well as far field blast effects. All this is possible for various scenarios, defined according to parameters such as wind direction and speed, gas composition and leak rate, leak and ignition locations, etc. Plant geometries can be imported directly from CAD systems, and results can be presented by a number of variables.

The main contractor for Flacscom was British Petroleum (BP) research and engineering group, though the company that has commercialised the FLACS code is Gexcon, a consultancy spin-off from Christian Michelsen Research (CMR, www.cmr.no), established specifically for this purpose in Bergen, Norway, in 1998. From the outset Flacscom wanted to make available to the petrochemical industry state-of-the-art gas explosion prevention technology, and to speed up the process of commercialisation of FLACS by rigorous testing (beta testing) by target users, and detailed review by ICT engineers. The main part of this research was to generate experimental data and theoretical knowledge needed to simulate gas explosion risk and to minimise the risk of fire escalation and structural failures.

> FLACS is an advanced tool for modeling of ventilation, gas dispersion, vapor cloud explosions and blast in complex process areas. FLACS is used for the quantification and management of explosion risks in the offshore petroleum industry and onshore chemical industries. The development of FLACS has been carried out with the full cooperation, support, direction and funding of 10 international oil and gas companies; BP, Conoco, Elf, Exxon, Gasunie, Gaz de France, Mobil, Norsk Hydro, Phillips, Statoil and three legislative bodies, Health & Safety Executive, Norwegian Petroleum Directorate and BMFT (Germany). The former recognised the need to have a tool that would enable them to design process areas and offshore platforms in order to minimise the risk from gas explosions. The steering committee for the Gas Safety Programmes has through 16 years of development ensured the compatibility of the code with their practical requirements.
>
> (www.gexcon.com, accessed 11 April 2004)

According to *Prosoma*, the main market for the FLACS code consists of the petrochemical companies themselves, and the many specialist engineering consultancy firms that support the sector. Other potential customers include the communities of process engineers and insurance company experts.

> We have got the first research tool which we use to improve and enhance the safety design of our facilities. The industry as a whole is dependent

> on its successful dissemination, adoption and utilisation. Because one thing we recognise is that in the area of safety it is the performance of the industry that is important, not of any individual company. For example, the Piper Alpha accident in 1987 cost 167 lives, and loss of facilities alone was over 2 billion pounds. What people could see at the time was not Piper Alpha, but the whole industry. If something like this happened again the legislation would get tighter and we would have spend a lot of money on research together with other companies, so if we show in practice that we care the government will not need to intervene with tighter legislation. The other reason we applied for Esprit funding is that we also wanted some expertise from outside our immediate group to audit what we did and independent auditors to look at how the work progressed against some sort of international standards.
>
> (BP project leader)

The network of formal partners for Flacscom included, apart from BP and CMR, oil and gas companies – Elf, Gaz de France, and Statoil from Norway – and a German consulting company, Inburex (www.inburex.com), an independent international company providing services in all fields of industrial and process safety including explosion prevention and protection, fire and chemical process safety and safety management. Graph 4.7 shows the personal network of the Flacscom project leader intertwined with personal networks of other key nominated individuals: important sources of information for innovation in relation to this project. Individuals and relational data are represented by the same conventions as above.

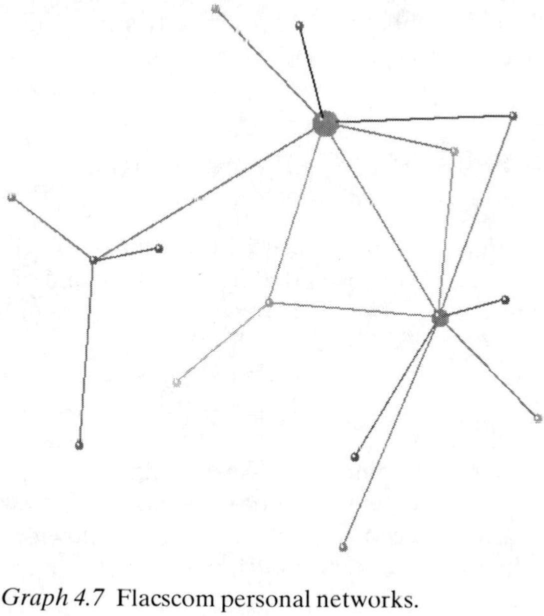

Graph 4.7 Flacscom personal networks.

The most central individual in the informal network is the BP project leader at the top of Graph 4.7, followed by the engineer/project manager at CMR at the bottom right-hand part of the graph, who worked together in the industry for more than a decade doing research in gas explosion technology. In Graph 4.7 there are also three cliques connecting BP and CMR with Gexcon, Elf and Gaz de France. As expected, few of the other nominations come from within the consortium, and include people from Inburex and Statoil, other external players such as Norsk Hydro and Sintef in Norway, and RisX in Aberdeen, Scotland. All the latter nominations, however, are isolates in Graph 4.7. Surprisingly only one of these individuals comes from the US Exxon–Mobil at Houston, Texas.

> One of the reasons why we have not involved colleagues from the United States more than we do is restriction on the Esprit budget. I thought it would create an unfair environment in that one half of the people would have cost covered, the other half would not ... But again I think the US colleagues were actually given our reports and at the end of each stage we had to have all our documentation delivered and they were actually invited to make suggestions. On some occasions they actually gave feedback. But in my experience when you work in a project which costs a lot of money and timescale is very short, always under pressure to deliver, you just close yourself down and concentrate.
>
> (BP project leader)

> I provided little input to the project, and all that input was provided without direct input from others. My contact at CMR was an indirect source of information, i.e. I learned about the general area from him but didn't consult with him about this project.
>
> (Exxon manager)

4.9 Parallel Electromagnetic Problem Solving Environment (Pepse)

Pepse was part of Europort 2, a large 26-month Esprit III project from 1994 to 1996. Europort 2 brought together 10 consortia that had the main aim of increasing awareness and building confidence in the effectiveness of parallel high-performance computing (HPC) platforms in such diverse industrial application areas as electromagnetic simulations, computational chemistry, cartoon animation, drug design, radiotherapy, traffic flow, earth observation and oil reservoir engineering. The main contractor for the Europort 2 project as a whole was Smith System Engineering (www.smithsys.co.uk) based at the Surrey Research Park in the south-east of England. From the outset the results of Europort 2 attracted considerable attention worldwide as can be seen from the following press release from Primeur – the premier HPC news service in Europe supported by the EC:

> Smith System Engineering is organising a three day seminar in Kobe, Japan, 19–21 May 1997, to present the results of Europort 2 and related work under the European Commission's Esprit programme. In the Europort project, 38 large serial commercial and in-house codes were ported to parallel supercomputers. This demonstrates the maturity of European industry in parallel HPC. The parallel codes ported and the software tools are now commercially available and are supported on a range of supercomputer systems.
>
> (www.hoise.com/CEC/AE-PR-03-97-15.html, 14 May 2004)

The Pepse project was a major step forward in HPC electromagnetic simulation within the aerospace industry. The software code (EMA3D) was initially produced by a US company EMA (www.electromagneticapplications.com) in Lakeside, Colorado, which provided the three-dimensional numerical solution of Maxwell's equations, and FEGS, now TransenData (www.transcendata.com) in Cambridge, England, which produced the graphical pre- and post-processor visualisation tool. The lead user and technical contacts for Pepse were at the military aircraft division of British Aerospace Defense (BAe) which also co-owned the EMA3D code. According to the Pepse entry in *Prosoma*:

> Given that aerospace systems increasingly depend upon digital signaling systems for flight control and systems operation, it is vital that potential sources of electromagnetic interference, such as lightning strikes, are anticipated and their effects assessed and protected against. Comprehensive physical testing is very costly both in time and money, and therefore it is important to ensure that it is effective and reduced to a minimum. Computational electromagnetic analysis can satisfy this need, by enabling examination of a wider range of cases than could be achieved through physical testing alone.
>
> Yes we originally got the EMA3D code three to four years prior to Pepse. BAe was interested in the intellectual property rights as well as the knowledge EMA developed on electromagnetic simulations ... but I mean we couldn't just take their code, develop it and then sell it, so we needed to cooperate with them and they were quite happy to get further involved within the Pepse consortium.
>
> (BAe manager)

Moreover, according to the Pepse's technical leader at BAe, the 'parallelised' code provided not only reductions in the overall design cycle time but also improved quality of visualisation results and technical innovation that extended much beyond the aerospace industry in areas such as the effects of electromagnetic fields from mobile phones on human brains and the discharge of static electricity across ICs:

> I would say the main innovation is the parallelisation of not just the electromagnetic solver, EMA3D, but also the pre and post processor file. This provides a reduction of the overall design cycle time and total cost in the practical industrial conditions. Furthermore, the ability of the code to model esoteric materials and wires and the use of perfectly matched layers, allows the code to be used to model a number of EM scenarios, including electromagnetic compatibility, electromagnetic pulses antenna design and stealth in a number of diverse design areas, such as civil and military aircraft, automobile, biological, anechoic chambers, microelectronics and the railway.
>
> (BAe manager)

The formal partner for the parellelisation of the serial code for Pepse was an interdisciplinary research centre CERFACS (European Centre for Research and Advanced Training in Scientific Computation) near Toulouse, France. CERFACS develops advanced methods for numerical simulations and algorithmic solutions of large scientific and technological problems of interest for research as well as industry, requiring access to the most powerful supercomputers presently available. It has five shareholders: CNES (the French Space Agency), EADS France (the European Aeronautic and Defence Space Co.), EDF (Electricité de France), Météo-France (the French meteorological service) and SNECMA (Société Nationale d'Etude et de Construction de Moteurs d'Aviation). Approximately 100 people work at CERFACS, including, in early 2004, 90 researchers and engineers from 10 different countries. They work on specific projects in six main research areas: parallel algorithms, aerodynamics, combustion, climate and environment, data assimilation and electromagnetism.

> Our link with CERFACS goes back to 1989. They sent out, I think originally, it was a letter to the directors explaining what they were trying to do, having an open day and invited people from select organisations and the European Union. I got sent there and this was nothing to do with Esprit. And then when Pepse came along 4 years later, I knew who was interested in this kind of applications and gave them a ring.
>
> (BAe manager)

The Pepse project network also included the following formal academic partners: the Department of Electrical and Electronic Engineering in Bradford University which had already been a user of the EMA3D serial code since the beginning of the 1990s for non-military electromagnetic applications, and the Department of Civil Engineering at the University of Wales at Swansea. Graph 4.8 shows the personal network of the Pepse project leader intertwined with the personal networks of other key people nominated as important sources of information for innovation in relation to the project. Individuals and nomination relational data are represented by the same conventions as above.

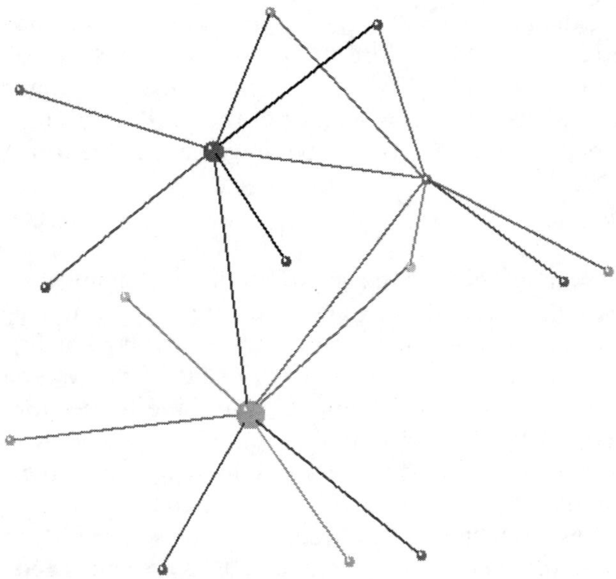

Graph 4.8 Pepse personal networks.

The most central individual in the network is not the technical leader from BAe but a professor at the Department of Electrical and Electronic Engineering, University of Bradford, the largest sphere in the middle bottom of the graph, who has published widely in the field of applied electromagnetics. The BAe engineer (upper-middle sphere in the graph) is the second most central individual in the network who nominated the professor at Bradford in the first place. Moreover, an interesting finding in Graph 4.8 is that this informal network is highly cohesive in the United Kingdom, having four cliques connecting BAe with Smith Systems Engineering, FEGS and the University of Bradford within the project, and also, outside the formal project network, with MIRA a user company of electrical and electronic systems at Nuneaton. But the other project partners and nominated engineers who are based in the United States (EMA) and France (CERFACS) are isolates.

4.10 Telematics programme dissemination: Project Information Prepared for Exploitation and Reference (PIPER)

Piper was a three-year fourth Framework project from 1996 to 1998 within the Telematics Applications Programme (TAP) to support actions for services of public interest. The project aimed to produce a software tool-kit for enabling publication on the web (and CD-ROM) of several hundreds of TAP deliverables originally formatted as Microsoft Word (Rich Text Format,

RTF) documents. It is worth pointing out here that Microsoft Word did not provide the option to save documents as HTML files in the mid-1990s. The idea at the heart of the project was therefore to easily transform RTF to HTML files and capitalise on the ubiquity of the web for disseminating widely and free of charge research results from part of the TAP supported by the EC. The latest version of the tool kit can be downloaded free from http://piper.ntua.gr.

According to *Prosoma*:

> The Tool Kit helps disseminators, working on projects in the Administration and Urban & Rural sectors of the EC's Telematics Applications Programme, by converting project deliverables ready for publication on the web. The Tool Kit prepares HTML and JPEG files from Word documents. It presents each and every deliverable in three frames for ease of navigation: Title, Table of Contents, and Content. The Tool Kit incorporates hidden information – meta data – in the deliverable, making it more likely that it is appropriately indexed and retrieved by Internet Search Engines. Font styles and paragraph alignments used in the Word document are retained after conversion; URLs are converted to active links; and links to footnotes and endnotes are automatically included.
>
> The Tool Kit is specifically designed to meet the needs of projects in the Administration and Urban & Rural sectors of the TAP. However all projects in the TAP can benefit from Piper's results. In fact anyone who wants to publish a large written document on the web will find that the Tool Kit offers advantages not available in commercial HTML converters. Any written deliverable, which is prepared to a set of EC guidelines, can be prepared for web dissemination with the Tool Kit. The Tool Kit also incorporates the EC guidelines and will check a document for conformance to these guidelines prior to conversion. A non-conforming document can still be converted – a log is produced that tells you how your document differs from a conforming one.

Moreover, according to Cordis, web-based dissemination of TAP deliverables would facilitate better exploitation of results, as without the use of appropriate indexing methods target audiences might never discover useful results. Also the preparation of text and graphics in the required formats for web publication was time-consuming and fiddly at the time and the final layout of the web files could be poor, without proper standardisation, making them less comprehensible to target audiences. By using the Piper tool kit, TAP projects had the potential to speed up the preparation of their deliverables into a consistent EC-approved layout which was readily indexed for the web.

The main contractor for the project was Level 7, currently ECSOFT (www.ecsoft.co.uk), headquartered in London, England. Other formal

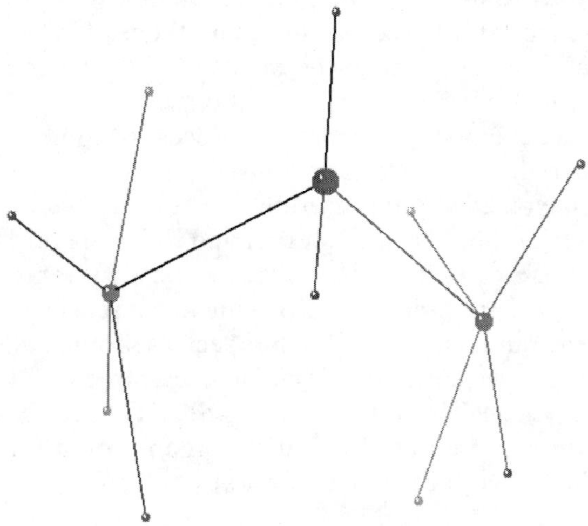

Graph 4.9 Piper personal networks.

partners were International Computers Ltd (ICL) in the United Kingdom, currently Fujitsu Consulting, and the Telecommunications Laboratory of the Electrical Engineering and Computer Science Department at the National Technical University of Athens (NTUA), Greece. Graph 4.9 shows the personal network of the Piper project leader linked with a few other key nominated individuals who were important sources of information for innovation in this project. Individuals and relational data are represented by the same conventions as above.

The most central individual in the network is the main UK contractor at ECSOFT, who has nominated people at ICL, the Telecom Lab at the NTUA, and also contacts in France and Belgium. It is worth pointing out that there is no cohesive group in Graph 4.9, though nomination rounds bring up people in the United Kingdom (e.g. University of Bradford), France (e.g. Euritis, Thomson) and Spain (e.g. UPC). Three EC officials/project officers based in Brussels are also nominated as key sources of information for the project. However, all these nominations are isolates in Graph 4.9, with the notable exception of the project leader who maintains reciprocal links with both the engineers at the Euritis and Telecom Lab in NTUA.

4.11 Intellectual Multimedia Property Rights Model and Terminology for Universal Reference (Imprimatur)

Imprimatur was a 39-month Esprit IV project carried out from 1995 to 1999. It aimed to build consensus on electronic copyright management and intellectual property rights (IPR) protection in the late 1990s and beyond. According to *Prosoma*, special interest groups (SIG) and major international consensus fora were convened to address a broad range of technical, legal, business and social issues to help establish a consensus on methods of information trading and digital transmission to meet the needs of IPR owners, intermediaries and end users from both sides of the Atlantic.

Concurrently technology was developed to model this process of consensus building surrounding digital rights for multimedia content, as the Internet grew to the biggest copy machine on earth, empowering users to copy with a few clicks vast amounts of material. Because the Internet and web disregard national boundaries, problems are caused by differences in legal systems, national cultures and so on for content industries. To achieve consensus in such infrastructure issues, a large number of stakeholders such as producers, distributors and consumers must be consulted. The main contractor of Imprimatur was the Authors' Licensing and Collecting Society (ALCS, www.alcs.co.uk), an internationally renowned British organisation for protecting IPR and in particular copyright based in London.

> The Authors' Licensing and Collecting Society Limited (ALCS) is the UK rights management society for all writers. Its principal business is to collect and distribute fees to writers whose works have been copied, broadcast or recorded. It also collects monies for lending and rental. ALCS was set up in 1977 in the wake of the campaign to establish Public Lending Right in the United Kingdom to help writers protect and exploit their collective rights ... ALCS is internationally recognised as a leading authority on copyright matters. It maintains a watching brief on issues affecting writers both in the United Kingdom and overseas, making representations to UK government authorities and the European Union.
>
> (www.alcs.co.uk, accessed 23 January 2003)

According to *Prosoma*, Cordis and the Imprimatur Report (1999) the project network also included the following formal partners: the British Library, CISAC (Confédération Internationale des Sociétés d'Auteurs et Compositeurs, France), Bertelsmann (Germany), Thomson Broadcast Systems (France), the Digital Copyright Forum (United States), EUSIDIC (European Association of Information Services, Luxembourg), IFPI (International Federation of the Phonographic Industry), MCOS (Music Copyright Operating Services), Croft Communication Consultants (United Kingdom), Tagish (United Kingdom), Teles (Germany), Telia (Sweden), plus the universities of Amsterdam and Florence; and as affiliate partners AEPO (Association of European Performers' Organisations), EDItEUR (Pan-European Book Sector EDI Group), EBLIDA (European Bureau of Library, Information and Documentation Associations), International STM (scientific technical and medical publishers), IFRRO (International Federation of Reproduction Rights Organisations), and the US Copyright Office.

> The Imprimatur consortium is trying to build consensus around digital rights trading. That sounds very easy. It isn't. At the moment, most content is sold in books, CD ROMs, videos and so forth. When this content migrates onto networks, the question is how can you trade it securely and fairly between the creator, the producer, the distributor and the consumer.
>
> (Imprimatur manager)

Graph 4.10 shows the personal network of the Imprimatur main contractor/project leader. The same conventions apply as above for representing individuals and ties showing nomination network data. The most central individual in the network is the main UK contractors themselves (the largest sphere at the upper right-hand side of Graph 4.10). However, what is even more interesting is that part of their network includes sources of information essential to this project not only in the United States (e.g. Digital Copyright Forum, Copyright Clearance Center), but also as far away as Australia. The network also includes sources in Scandinavian countries and the Netherlands.

Again, it is notable how nominated sources outside the Esprit project itself nominate sources of information within the project so that networks which might have been thought to have been internal to Esprit are in fact intertwined with external information networks. The extent of overlap can be seen in the case of an American contact (bottom right-hand side of the graph) from the Copyright Clearance Center who is linked with the main UK contractors, but also with another two of their nominations: a professor at a Dutch university and an ALCS manager. Such overlaps allow valuable information for Esprit innovation to flow back and forth from the United

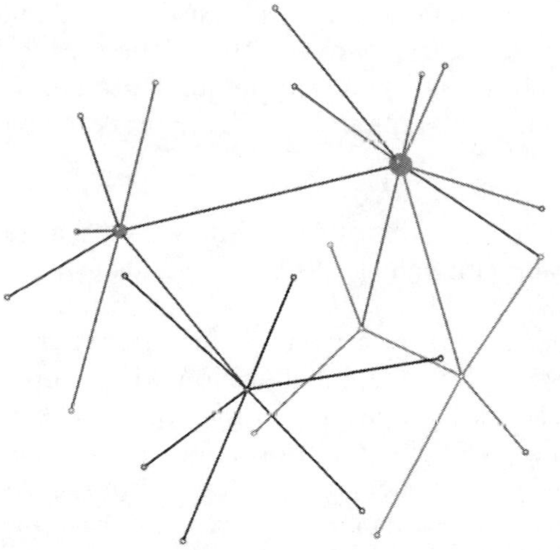

Graph 4.10 Imprimatur personal networks.

Kingdom to the United States via a number of direct and indirect ties within and outside the ALCS.

> Let me show you something . . . this is a database of contacts I have made 768 people on it all over the world. Put Australia in . . . these are all the people I know and have worked with, 36 people in Australia. Put USA in . . . you will get all people I know personally . . . you get 71 in the USA.
>
> Look at our consensus forums, we had a very regular re-attendance at our forums, altogether we had about 260 people, actually no we had more . . . about 275 because we ran five in Europe and one in the United States. The one in the United States was slightly different and I think certainly created a community of people who were dedicated, trying to move forward in this way, and to understand the issues and to take a non confrontational approach to them. Because an essential part of what we were doing was trying to create a non-confrontational space in which these issues could be debated confidentially and freely.
>
> (Imprimatur manager)

Note that there are only two cliques in Graph 4.10 connecting the Imprimatur project manager with a close friend and colleague within ALCS, the Copyright Clearance Center near Boston in the United States and the Institute for Information Law at the University of Amsterdam in the Netherlands. It seems that trust and mutual interest hold these information networks together. Neither is easy to establish and both take time and effort. A concern encountered frequently among those interviewed was that the EC was insufficiently sensitive to these arrangements and to the personal investment that had gone into making them. In forcing contacts outside their own personal networks on those working on Esprit projects, the EC put at risk these Esprit experts' personal information networks. Consequently, the EC endangered the very innovation it was trying to encourage. The concern expressed by the technological leader of the main Imprimatur contractor is typical of other cases (see, for example, the quote from the main Amulet2 ARM contractor below).

> My network of contacts spans the world, reflecting the global nature of IPR. It also spans private companies, NGOs, INGOs, supra-governmental organisations like the UN and OECD and governments themselves. One extremely irksome thing the Commission often tries to force on those who work in Esprit is the collaboration with people outside this network of contacts. Such people are outside my network of contacts for both personal and professional reasons. Therefore when the EC insists one works outside one's network, such collaboration is bound to fail because it is neither based on mutual interest nor trust.
>
> The reason the Commission has to impose some partners is that they will be left out if they don't, and they put money into the pot in Europe,

> and occasionally they are saying why don't you pick up this company in trouble ... Yeah, all right we will have them in the project ... It is a pain but we did it because it helps. ... The EC is full of politics. Full of it, and we try and avoid that, and try and focus rather hard on what we try to do.
> (ARM manager)

Unsurprisingly, there seems to be some tension between project officers in the EC and participants in Esprit projects. Individuals interviewed insisted that their information networks are deliberate constructs that can easily be damaged by the brokering efforts of the EC to create its own dedicated networks. This happened in particular in the late 1980s.

> In the early days of Esprit II, XX was instrumental in arranging consortia. In those days a couple of phone calls would do it ... In 1989, Portugal and Spain had to be 'pushed in' (these countries became eligible for Esprit II) but it is difficult to have forced marriages between companies who have not chosen each other. Later such arrangements became much more difficult to make in this way. Over time, people got to know each other and who was worth working with. You knew somebody who knew somebody else. Contact led to contact and eventually to a professor at the University of Barcelona. It took a year to put together a project.
> (EC Esprit project officer)

> During Esprit II the project officers had about seven projects to take care of and during Esprit IV we have seen a large increase up to twenty projects or more. It is impossible to try to build informal links for all projects. My relationships are limited to the project manager and I have fewer contacts with others. One also tends to focus on problematic projects and less on those that function well.
> (EC Esprit project officer)

4.12 Evaluation of findings

Figure 4.2 highlights the significance of informal contacts in ICT innovation above. It summarises some of the main findings according to the relational nature of nominations with respect to project and organisational boundaries. It is worth noting that a nomination link shows that information considered of significant value for ICT innovation was exchanged between two individuals involved in one of the 10 RTD projects above. Formal links, i.e. within an Esprit project network and formal agreement with the EC, are the lightest grey when they connect people within the same organisation and darkest grey when the two individuals belong to different organisations. Informal links, i.e. connecting two individuals who are not both members of a formal Esprit project consortium, are black when they link people from two organisations and lightest grey/mid-grey when they link people who belong to the same organisation.

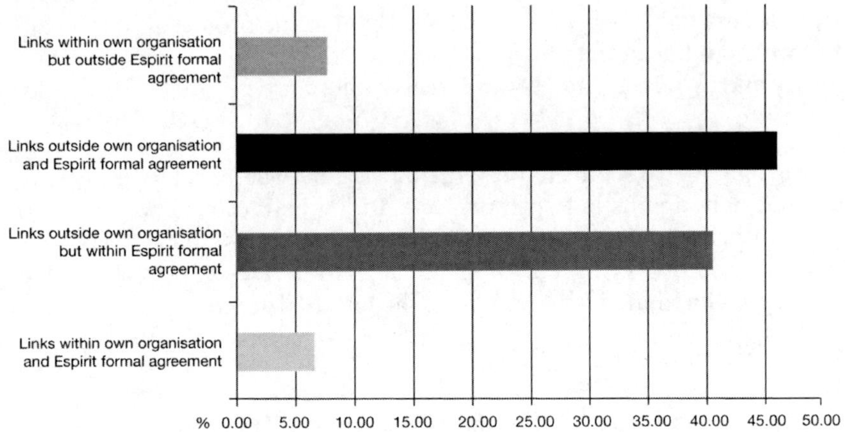

Figure 4.2 Formal vs. informal links for 10 Esprit RTD projects.

It is striking that more dyadic links valued for innovation across the 10 projects are informal (53.1 per cent) than formal (46.9 per cent), highlighting the significance of exchanging information valuable for innovation across project boundaries with personal contacts not tied in any formal contractual manner with the EC. Moreover, the differentiation within formal (darkest grey 40.6 per cent and lightest grey 6.3 per cent) and informal (black 46.1 per cent and orange 7.0 per cent) categories highlights even more clearly the importance of information exchange across organisational boundaries since very few links (lightest grey/mid-grey) are between people from the same organisation.

> The new world of IST research is one of personal networks (or network individualism, see, also, Wellman 2002) in which the role of the individual is crucial. Information flow in these networks takes place between individuals, not organisations. This is far removed from the notion of organisational research originally envisaged in Esprit. Research on informal information flows in Esprit finds that the informality which the IST Programme seeks is indeed present and important in ICT innovation. ICT innovation centres round the individual rather than organisation. The networks of individuals extend far beyond formal Esprit collaborations. In some contrast to this informal information flow through personal networks, the formal organisation of Esprit collaborations may appear, at least to those outside Esprit, inward-looking and self-perpetuating.
>
> (Assimakopoulos *et al.* 2000: 20)

Given the level of continuing support – more than €5.5 billion spent for Esprit research from 1983 to 1999 – that formal partners get from the EC through RTD programmes such as IST, the findings of Figure 4.2 have significant implications for how the EC is going to accommodate and institutionalise these key sources of information for innovation from much beyond project boundaries. As it has been suggested in Assimakopoulos *et al.* (2000):

> It is no longer necessary to see the project as the fundamental unit of Esprit and IST organisation. Networks and clusters of projects break down project boundaries. These alternative units of organisation are also appropriate to the wider dissemination of results and benefits of Esprit and IST research, and permit the exploitation of complementarities and synergies. Can these new units of organisation be managed in the same way as the old? What is the new role of the project officer when the project is no longer central to IST research activities? How does the project officer control IST research when so much of the input comes from outside IST, and so much of the benefit goes beyond IST?
>
> Esprit has long managed research by and through organisations. It has recognised only the organisational entity. And yet research is conducted by individuals. In as much as research is the product of individuals working in organisations rather than organisations employing individuals, the role of the individual must be recognised and accommodated. Should IST manage individuals rather than simply organisations?

As is currently evident in the IST programme of the sixth Framework (2003–6) supported by the EC, the new funding instruments of 'integrated projects' and 'networks of excellence' support research consortia organised in knowledge clusters and innovation networks encompassing a much broader consortium of partners from within and also outside the European Union. Specific agreements for international cooperation, for example with the US National Science Foundation (NSF), have also strengthened the possibility of cooperation with partners across the Atlantic who can apply for their portion of funding from the NSF as their counter-partners do from the EC. The role of key individuals, however, has yet to be properly addressed as the EC keeps a strong emphasis on legal entities, such as organisations, for satisfying the accounting exigencies of a large bureaucracy.

> Esprit contracts are with organisations, and yet research, particularly high technology research, is conducted by individuals, many of whom are exploiting their personal intellectual and social capital. In the academic world, research grants are awarded to individuals to be administered by their employing organisation. In as much as the role of individuals is critical in high technology innovation, this system should be adopted for the IST programme in the sixth Framework and beyond.
>
> (Assimakopoulos *et al.* 2000: 21)

Furthermore, Table 4.2 presents the main findings according to a simple 'developed/less favoured' classification of countries involved in the 10 cases. 'Developed' includes the following countries: the United Kingdom, Ireland, France, Belgium, Germany, the Netherlands, Austria, Italy, Finland, Norway, Sweden, the United States and Australia. The underlying assumption here is that all 'developed' countries encompass organisations and individuals embedded in well-developed and information-rich regions. For example, nominated contacts in Italy are much more likely to be found in the Milan metropolitan region than in the 'less favoured' regions in the south of the country. On the other hand, 'less favoured' countries include Greece and Spain where regions have received continuous support from EC instruments such as the 'structural funds' for regional development, or Brazil, which is wholly part of the developing world.

As Table 4.2 shows, the information flows of only three of the 10 projects connected people from the developed (D) countries with people in the less favoured (LF) countries of the south. Two of these projects developed software for manufacturing and the utilities and the third one focused on the dissemination of research results from the Telecoms Applications programme. It is notable that information flows between developed and less favoured countries with respect to application areas in high-velocity markets in ICT hardware are absent, as is software development in services (e-banking), aerospace and defence. Out of the 172 dyadic links most valued for IT innovation, the vast majority (88 per cent) were confined within the information- and knowledge-rich countries in northern Europe, the United States and Australia.

Since the European Union placed a premium on supporting less favoured regions in the south for all Framework programmes from the early 1980s to the early 2000s, this is an important finding. It reflects the discrepancy between the ability to provide financial support through instruments such as 'structural funds' using formal contracts and the 'failure' of policy to re-direct personal information flows from the information-rich in the north to

Table 4.2 Analysis of links between developed and less favoured (D–LF) countries

Project	*D–D (%)*	*D–LF (%)*	*Total number of links*
Amulet	10 (100)	0	10
Delphi	21 (100)	0	21
Improve	21 (100)	0	21
Fires	8 (50)	8 (50)	16
E2S	17 (100)	0	17
Timely	5 (45)	6 (55)	11
Flacscom	18 (100)	0	18
Pepse	19 (100)	0	19
Piper	5 (42)	7 (58)	12
Imprimatur	27 (100)	0	27
Total number of links (%)	151 (88)	21 (12)	172

the information-poor in the south. Furthermore, the nomination of people as far apart as Norway, the United States and Australia indicates the global nature of IT innovation networks, as main UK contractors accommodated informal, unacknowledged partners outside the European Union with the aim of acquiring information valuable for their innovation from far beyond the boundaries of the European Union (see also the more detailed analysis of internal and external linkages below).

> At least [the Greek partner] is being exposed to the way a network works in a project ... the previous position was they had nothing, no network, and they were isolated. Are they better off in a project that has a separate network or not? I would say yes they are because they can watch how a network works.
>
> (ARM manager)

Graph 4.11 shows the global innovation network of main UK contractors based on the 10 projects and the 16 nominated countries (Assimakopoulos and Macdonald 2003). Countries are represented by spheres positioned in a three-dimensional space according to their structural equivalence in the network (Burt 1987; see also the Appendix). The size of spheres varies according to their degree centrality measures (Wasserman and Faust 1994: 178; see also the Appendix). As was expected, the most central country in the network of Graph 4.11 is the United Kingdom itself. However, what is surprising is that the United States is the second most central country in Graph 4.11.

The United Kingdom and the United States are followed in third, fourth and fifth places by France, Germany and Belgium. The centrality score of the United States, a non-EU country that was not allowed to participate as an equal partner in EU-funded RTD projects of the third and fourth Framework programmes, throws new light on the role of informal partners in Esprit innovation networks. To this end, the patterns of internal and external linkages of the 10 projects are further explored in Table 4.3.

Table 4.3 summarises the main findings for internal linkages (dyadic ties within the boundaries of the European Union) and external linkages in the 10 projects. Note that most external linkages were dyadic ties connecting individuals in an EU and a non-EU country, while in some cases both individuals worked for organisations outside the European Union. As Table 4.3 shows, in only two of the 10 projects were the information flows confined to the European Union. Of the 172 dyadic ties, almost half (43 per cent) transcended the boundaries of the European Union. This is an important finding, given that none of the 10 projects had any formal partners outside the European Union. If there was no contractual need to involve outsiders, it seems that the only plausible explanation for these external links is that individuals in the large majority of projects believed that their own external contacts were particularly useful for innovation (Aldrich and von Glinow

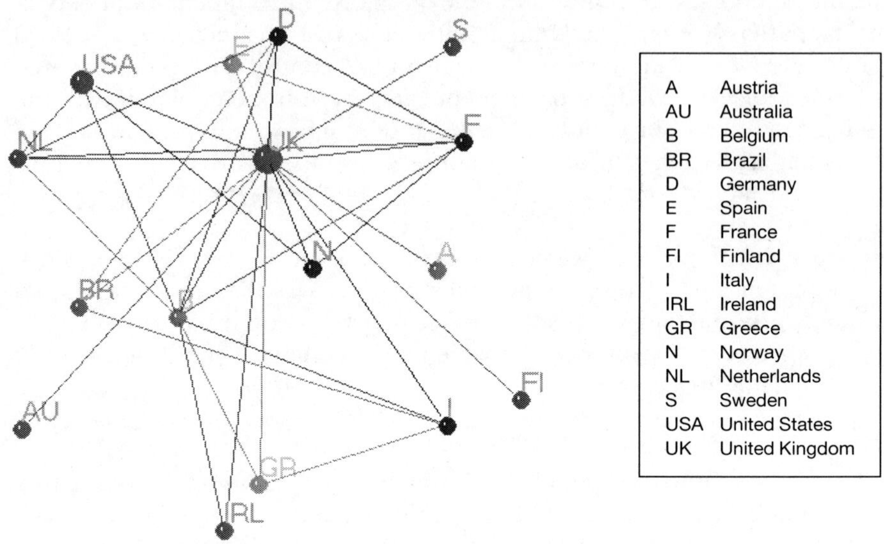

Graph 4.11 A country-based network analysis of links for 10 RTD projects.

Table 4.3 Internal vs. external links for 10 RTD projects

Project	*Internal links Number (%)*	*External links Number (%)*	*Total number of links*
Amulet	10 (100)	0	10
Delphi	15 (71)	6 (29)	21
Improve	21 (100)	0	21
Fires	9 (56)	7 (44)	16
E2S	3 (18)	14 (82)	17
Flacscom	3 (17)	15 (83)	18
Pepse	8 (42)	11 (58)	19
Piper	6 (50)	6 (50)	12
Timely	4 (36)	7 (64)	11
Imprimatur	19 (70)	8 (30)	27
Total number of links (%)	98 (57)	74 (43)	172

1992). It would seem that the majority of Esprit projects with main UK contractors accommodated informal, unacknowledged partners outside the European Union with the aim of acquiring information valuable for their innovation despite a Euro-centric RTD innovation policy promoted by the EC.

As might have been expected, the majority (58 per cent) of the external linkages of the main UK contractors were with the United States. EU firms have generally been eager to collaborate with the US companies because of their technological lead in ICT (Narula 1999). The linguistic and cultural connections of individuals in UK firms and other institutions also explain US dominance of their external linkages. It has long been known that UK organisations participating in the EC's RTD programmes have more collaborative links than their partners (Georghiou 1992). It is worth speculating that the attraction of a UK partner may lie less in its intrinsic qualities than in its links with the United States (Coles, Harris and Dickson 2003). Also striking is the global spread of external linkages: through these individuals, main UK contractors maintained important links with colleagues in countries as far apart as Australia and Norway.

4.13 Summary

The Appendix (p. 200) discusses some of the methodological issues related to mapping personal innovation networks and collecting and analysing relational quantitative and qualitative data for the 10 case studies discussed above (Wasserman and Faust 1994). Both the Cordis and Prosoma databases were used to identify a convenience sample (Bryman 2004) of 67 RTD projects with the main UK contractor supported by the EC and its Esprit III and IV programmes throughout the 1990s. The 10 projects that returned nomination data for more than two rounds of fieldwork provided Ptolemaic views and insights into the main UK contractors' egocentric networks within Europe and worldwide (Wellman 1988). The analysis of the main findings and discussion also made use of different network measures such as centrality and cohesiveness. A rigorous framework based on three-dimensional scaling using Euclidian distances as a measure of structural equivalence (see Appendix) facilitated comparisons across all 10 cases, also providing insights with regard to the value of informal, personal linkages in ICT innovation that connected key players in the United Kingdom with their counterparts in Europe, the United States and worldwide.

In Sections 4.2 to 4.11 each RTD project was presented with regard to its broad aims and objectives, main contractor and formal network supported by the EC. The first three projects, Amulet, Delphi and Improve, dealt with hardware-development-related issues; the next five projects, Fires, E2S, Timely, Flacscom and Pepse, dealt with software development; and the last two projects, Piper and Imprimatur, dealt with soft issues such as building consensus for trading multi-media property rights on the web. The analysis equally focused on the informal networks of main UK contractors, highlighting in particular the inter-dependencies between formal and informal personal networks for ICT innovation. Regardless of whether projects developed innovative world-class hardware, software solutions ranging from manufacturing to services to aerospace and defence, or focused on soft issues, the main findings draw attention to the significance of personal unanticipated contacts across organisational boundaries worldwide.

As Figure 4.2 shows, more inter-personal links were informal than formal in nature, connecting people much beyond the boundaries of formal consortia, based in the information-rich and developed countries of northern Europe, the United States and Australia (see Table 4.2 and Graph 4.11). Furthermore, external linkages across the boundaries of the European Union were present in eight out of the 10 cases presented above, highlighting the value of partners from the United States and other places worldwide (see Table 4.3). The comparative evaluation of findings also adds validity to the claim that EC policy for ICT innovation should shift away from creating a 'fortress Europe' and encompass key individuals and their informal networks worldwide in its current and future Framework programmes for RTD.

5 An astronomer's view of the origins of a national GIS community

5.1 Introduction

This chapter discusses the origins and early critical stages of development of a new technological community in a national scale (i.e. Greece) from the early 1980s to the mid-1990s. It adopts a network approach and focuses on a specific family of computerised technologies, geographic information systems (GIS) (Goodchild 1992). The underlying argument is that the evolution and dynamics of the GIS community in Greece can be analysed through the adoption, implementation and diffusion processes of GIS innovations in a broad range of settings over its early critical years from the early 1980s to the mid-1990s (Assimakopoulos 1996, 1997a, 1997b, 1997c, 2000).

In the past many network ethnographic studies of traditional communities were carried out in small and inward-looking social systems such as villages and monasteries where the boundaries were clear and the total population was known, making the identification and sampling of actors and networks easier (Sampson 1968; Wellman and Leighton 1979). Unlike those studies, a major challenge for this research was to draw the boundaries of the Greek GIS community and to identify the GIS actors who adopted and implemented GIS innovations across a broad range of institutional settings and disciplinary backgrounds throughout Greece. For this reason snowball sampling (see Chapter 1) and a multi-stage, multi-site approach was followed since it was not possible to identify the various stakeholders of the Greek GIS community at one location or in a single round of fieldwork.

Greece was a suitable laboratory to study the origins and development of a GIS community for two interrelated reasons. First, the research inevitably drew upon the personal experiences of the author who has in-depth knowledge of the Greek language and culture, and also had an opportunity to be involved as a researcher with GIS innovations at the universities of Patras in Greece and Sheffield in the United Kingdom, from the late 1980s until the late 1990s. Second, Greece is a small country (population 10 million in 1991) where the common language and culture coupled with the same system of government have allowed the development of a relatively small GIS community at a national scale. As a result it was practically possible to track down most of the individuals and other actors who were members of the Greek GIS community, unlike the situation in much bigger countries such as the United Kingdom. Overall I carried out more than a hundred in-depth interviews and participated in various GIS workshops and events where I updated and extended my list of contacts through the early 1990s, selecting the individuals, teams and organisations who most often came up as important to meet, either because of their formal positions and relations, or because of their experiences and knowledge of GIS technologies. Supporting evidence and triangulation were achieved mainly through the peers of these GIS experts.

This chapter is divided into seven additional sections. Section 5.2 identifies the main institutional groups at the macro-innovation 'triple helix' environment of government–academia–private sectors of the Greek GIS community. Section 5.3 highlights the essential complementarities of interests between the main institutional community components and the main technology components of hardware, GIS software, digital data, skills and other services. Section 5.4 focuses on the dynamics of the emergence of the Greek GIS community since its origins in the early 1980s to the formation of its critical mass (Rogers 1993, 2003) in the mid-1990s. It also evaluates the findings of the earlier discussion with particular reference to four stages in the development of this community: cosmopolites, innovators, early adopters and early majority.

Subsequently, in Sections 5.5 to 5.8, the essential complementarities of interest between community and technology components are explored with socio-metric techniques, SNA and visualisation software. Section 5.5 puts forward a graph of the critical mass of the Greek GIS community based on 51 teams and 190 reciprocal multi-stranded linkages. Then Section 5.6 sheds light on the prominence and centrality of GIS teams sharing particular backgrounds, such as the surveying engineering tradition of practice. Last, Sections 5.7 and 5.8 explore heterogeneity and discuss the institutional and disciplinary arrangements within and across relevant social groups and existing technological communities who handle geographic information and actively shape the development path of GIS technology throughout the country.

5.2 Institutional components of a new GIS community

About 50 GIS teams who came from various groups of institutions and made up the Greek GIS community were identified through their peers using a multi-site, multi-stage methodology in Assimakopoulos (1997a). Therefore the discussion that follows concentrates on the groups of individuals, or GIS teams, within organisations which are responsible for the adoption and implementation of GIS technology. GIS teams were chosen as units of analysis because in the large majority of organisations only a small group of individuals, rather than the wider organisation, is responsible for GIS adoption and implementation. Moreover, in many organisations such as universities several teams which usually work in different laboratories or departments may be involved.

Table 5.1 shows the 51 GIS teams who were identified as members of the Greek GIS community. It also shows the organisation in which each GIS team was based and the main activities of the team related to GIS technology. The five institutional groups (central government agencies, municipalities, utilities, universities and private firms) of GIS teams can be viewed as the functional components of the Greek GIS community within the triple helix of university–industry–government relationships (Etzkowitz and Leydesdorff 2000) enabling the adoption and diffusion of GIS innovations in a broad range of organisational settings, and a parallel shift from mode 1 to mode 2 knowledge production (Gibbons *et al.* 1994; see also Section 3.2) related to the handling and analysis of geographic information, and in particular GIS technology.

Three criteria were used to distinguish these 51 GIS teams from other groups using ICT in Greece. Primarily, the adoption and use of GIS software for the development of GIS applications was considered essential, since many organisations use and analyse geographic information without a GIS. Second, it was thought that GIS teams should also have experience and expertise with GIS adoption and implementation, and, third, it was considered important that the GIS work of individual members of a GIS team should be known to other individuals and groups of the Greek GIS community. The boundaries between Greek GIS teams and non-Greek GIS teams which have played a role in the development of the Greek GIS community are also clearly defined. Table 5.1 excludes foreign GIS teams, but includes Greek expatriates who have set up teams in Greece: for example, the Foundation for Research and Technology Hellas (FoRTH) at Heraclion in Crete.

The size of most of these GIS teams was generally small (less than five people) but varied according to the institutional setting and the organisational resources available for the adoption and implementation of GIS technology. Within the public sector, central government agencies such as the National Mapping and Cadastre Organisation and the Digital Cartography Department of the Hellenic Military Geographical Service had three or four senior staff/GIS experts and about 10 technicians for data handling and analysis. On the other hand, the GIS team at the Directorate of Urban Planning within the Ministry of Environment, Planning and Public Works consisted of a geologist and an architect/planner, plus one or two part-time support staff. Within the universities and the research institutes, most GIS teams consisted of one or two full-time academics who had a major interest in GIS supported by a small group of part-time research staff and graduate students who worked on the development of GIS applications (for example, the Laboratory of Spatial Planning at the University of Patras, or the Department of Environmental Studies at the University of the Aegean). However, larger size groups could be found at the Laboratories of

Table 5.1 Groups of institutions and GIS teams forming the Greek GIS community in 1994–5

Group of institution	*Organisation / Department*	*GIS team's main activity*
Central government		
Ministry of Defence	Hellenic Military Geographical Service	1. Digital mapping and data production, military projects
	Hellenic Navy Hydrographic Service	2. Digital mapping and data production, navy projects
Ministry of Environment, Planning and Public Works	National Cadastre and Mapping Organisation	3. Digital mapping and environmental monitoring
	Directorate of Urban Planning	4. Urban planning
	Directorate of Environment	5. Environmental management
Ministry of Agriculture	Directorate of Forests	6. Digital mapping and forest management
National Institute of Mineral and Geological Research	Information Technology (IT) Centre	7. Digital mapping and resource management
Municipalities		
Nikea City Council	Technical Services	8. Urban planning
Piraeus Metropolitan Area	IT Services	9. Urban and transportation planning
Athens City Council	Cleansing Services	10. Refuse management
Thessaloniki City Council	Topographic Department and IT Services	34. Digital mapping and cadastre
Patras City Council	Technical Services	41. Urban planning
Heraclion City Council	Technical Services	45. Urban planning
Volos City Council	GIS Department	46. Cadastre and urban planning
Rodhos City Council	Technical Services – Architecture	50. Urban planning
Kos City Council	GIS Department	51. Cadastre and urban planning
Utilities		
Hellenic Telecommunications Organisation	Radio Links	11. Visibility studies
Greek Public Gas Organisation	Gas Distribution Networks	12. Network management
Universities		
National Technical University of Athens	Rural and Surveying Engineering – Laboratory of Cartography	13. Cartography
	Rural and Surveying Engineering – Laboratory of Remote Sensing	14. Remote sensing
	Rural and Surveying Engineering – Laboratory of Photogrammetry	15. Photogrammetry
	Rural and Surveying Engineering – Laboratory of Geography	16. Geography and regional planning
	Architecture – Laboratory of Urban and Regional Planning	17. Urban planning
	Civil Engineering – Laboratory of Hydraulics and Maritime Engineering	18. Water engineering and hydraulics

Group of institution	*Organisation / Department*	*GIS team's main activity*
Pantium University of Social Sciences	Department of Urban and Regional Development	19. Urban and regional planning
Agricultural University of Athens	Soil Surveys and Mapping	20. Soil management
Aristotle University of Thessaloniki	Rural and Surveying Engineering – Laboratory of Cartography	35. Cartography and cadastre
	Rural and Surveying Engineering – Laboratory of Urban Planning	36. Municipal information systems
	Architecture – Laboratory of Urban and Regional Planning	37. Urban planning
	Agriculture – Laboratory of Remote Sensing	38. Remote sensing
	Forestry – Laboratory of Forest Management	39. Forest management
University of Patras	Civil Engineering – Laboratory of Spatial Planning	42. Urban and regional planning
	Computer Engineering – Computer Technology Institute	43. R&D, utilities and emergency management, cadastre
Foundation for Research and Technology at Heraclion, Crete	Institute of Applied and Computational Mathematics	44. Transportation and environmental modelling
University of Thessalia at Volos	Department Urban and Regional Planning	47. Urban and regional planning
University of the Aegean at Mytilene	Environmental Studies	48. Remote sensing, ecology
Private Sector		
Marathon Data Systems	Sales Department	21. ESRI's software
Intergraph Hellas	Sales Department	22. Intergraph's software
Infotop	Sales Department	23. Erdas' and Panterra's software
Space Hellas	Sales Department	24. Laserscan's software
Omas	Sales Department	25. MapInfo's software
Ambit	Sales Department	26. Star's software
Eratosthenis	Engineering Consulting Firm	27. Digital mapping and data production, cadastre, transportation planning
Geomatics	Engineering Consulting Firm	28. Environmental management, cadastre and urban planning
Geomet	Engineering Consulting Firm	29. Cadastre, remote sensing
Terra	Engineering Consulting Firm	30. R&D, environmental planning and digital data production
Gaia	Engineering Consulting Firm	31. Urban and environmental planning
Epsilon	Engineering Consulting Firm	32. Environmental and regional planning, digital data production
Infocharta	Engineering Consulting Firm	33. Transportation planning and digital data production
Infodim	Engineering Consulting Firm	40. Cadastre, tax assessment
Viosfairiki	Engineering Consulting Firm	49. Environmental management

Cartography at the National Technical University of Athens (NTUA) and the Aristotle University of Thessaloniki (AUT) where the academic teams consisted of a professor plus four other permanent full-time academic staff with a major interest in GIS. Within the private sector the GIS teams had up to three individuals playing the role of GIS vendor or/and consultant with the exception of a few firms, such as the engineering firm Eratosthenis, who had many GIS projects throughout the country and a larger GIS team of engineers and technicians.

It should be noted, however, that there were both overlaps and shifts in membership of some of these GIS teams over time. Overlapping membership occurred both in the same sector (see, for example, Marathon Data Systems, the leading GIS software vendor, and the private consulting firm Eratosthenis) and across different institutional groups (for example, the Laboratory of Water Resources in the Department of Civil Engineering at the NTUA and Epsilon consulting firm). Over time some key individuals also moved between GIS teams either within the same or across different sectors. Only a few highly skilled individuals had the opportunity to move between GIS teams within the same or different sectors.

It is also worth pointing out that most teams consisted of individuals with a variety of expertise and skills. In some cases they were part of a bigger organisation which provided continuity even when particular individuals moved between different organisations or sectors. In other cases they were dependent to a significant extent on the expertise and skills of particular individuals: for example, the founder and president of Marathon Data Systems (www.marathondata.gr), the leading ESRI (www.esri.com) vendor of popular GIS software such as ArcInfo in Greece. It should also be noted that some teams such as software vendors had a single interest in terms of GIS adoption and implementation whereas some individual academics had multiple areas of interest (for example, FoRTH and Omas/InfoCharta private firms).

Table 5.2 shows the number and percentage of teams that belonged to each of the main functional components of the Greek GIS community. From this it can be seen that 35 per cent of the teams identified came from academia. Academic teams formed the largest group in the Greek GIS community adopting and implementing GIS in a variety of institutional and disciplinary settings. Many of these academic teams (for example, the Departments of Topography and Geography, see, www.survey.ntua.gr at the NTUA) had been exposed to the leading edge of GIS-related theories and methods because of their connections with various networks abroad. It can be argued, therefore, using the terminology developed by Constant (1980), that as a result of knowledge creation and exchange processes about GIS innovations in the early critical days for GIS diffusion in Greece, and because academics were closer to the theoretical foundations of their respective scientific communities, these people were in an advantageous position to understand the first 'presumptive anomalies' (see Section 2.5) in terms of

Table 5.2 Functional components /institutional groups that make up the Greek GIS community

Institutional groups of GIS teams	*Number of teams*	*Percentage*
Central government	7	13.7
Municipalities	9	17.6
Utilities	2	3.9
Academia	18	35.3
Private sector	15	29.4
Total	51	100

computerised geographic information handling and analysis for a number of existing traditions of practice in Greece, including surveying engineering and spatial planning. Subsequently these individuals and academic teams drove radical technological change for existing disciplinary and professional groups in Greece and also played a leading role in the emergence of a new technological community and tradition of practice related to GIS technology in Greece throughout the late 1980s and early 1990s.

The next largest group of teams is those of GIS software vendors and engineering consulting firms. Private sector firms represented less than a third (29.4 per cent) of the total number of teams promoting not only the adoption of GIS software and hardware but also providing various specialised services in relation to geographic information handling and analysis in Greece. Some of these firms, such as Terra (www.terra.gr/eng/intro.htm) and Epsilon (www.epsilon.gr) were set up by GIS experts with a direct link to academic institutions such as the NTUA. As a result of their contacts with university laboratories, some individuals in these GIS teams had also under stood the 'presumptive anomalies' related to the handling of geographic information since the late 1980s.

For example, individuals like Dr Thanos Doganis, who carried out his PhD at the NTUA and worked for the engineering consulting firm Eratosthenis from the late 1980s up to 1993, co founded Terra at the science park of the National Research Centre 'Demokritos', developed GIS applications in various institutional settings and has been initiating radical GIS change for the existing surveying engineering tradition since the very early days of GIS diffusion in Greece. Unsurprisingly, in 2001 the same key individuals played a leading role in setting up the Geographic Information Society (www.hellasgi.gr), the Greek member of the European umbrella Organisation for Geographic Information (www.eurogi.org).

More than a third of the total number of GIS teams came from the public sector: central government agencies, municipalities and utility organisations. Teams based in municipalities represented the largest part of the Greek public sector with 17.6 per cent of the total. In contrast to the teams who came from central government, however, these teams have had a limited influence on the development of the Greek GIS community because of the small size of the authorities and the lack of resources available for GIS applications. Central government organisations, on the other hand, although they accounted for only 13.7 per cent of the total, form a significant group because many GIS applications and data sets developed in this group of institutions have had national importance. Last, utilities represented only 3.9 per cent of the total.

From the outset Tables 5.1 and 5.2 give a good indication of the size and composition of the Greek GIS community in the mid-1990s. However, it must be pointed out that these numbers are indicative, rather than definitive, because there were overlaps between some teams, and the boundary between GIS and other ICT applications was, and still is, not a hard and fast one.

5.3 Complementarities of GIS interests in the triple helix of university–industry–government relationships

Teams from different institutional settings have played complementary roles in the Greek GIS community. These GIS stakeholders interact and relate to each other not only because they share a common interest but also because they have to carry out responsibilities and tasks which are often essentially complementary in nature. The different types of relationships and the strength of the interdependencies between and within the various institutional and disciplinary groups are explored and visualised in Sections 5.5 to 5.8.

Table 5.3 presents the complementary areas of interest that the different stakeholders have in the Greek GIS community. It shows that people and groups who work with GIS within central and local government institutions and the utility companies are primarily interested in long-term digital data provision and supply services. There is a long tradition of collecting and analysing geographic information by Greek public sector institutions (for example, the Hellenic Military Geographical Service, www.gys.gr/english/EN3.htm), so it is expected that this group of GIS stakeholders will continue to do so, perhaps also adopting a more open and market-oriented strategy towards information sharing in the twenty-first century. Throughout the 1990s the interests of government agencies in data production were also complementary to those of the private sector since some GIS vendors and engineering consulting firms produced digital data for various private and public sector agencies (see, for example, Eratosthenis www.eranet.gr).

Table 5.3 also shows that academics have played a significant part in the development of GIS knowledge and skills in Greece through the organisation and delivery of GIS-related courses and training seminars (for example, the Department of Topography at the NTUA). They have also had an important role to play in raising awareness about GIS in Greece. Many of the university-based teams have carried out a number of research contracts

Table 5.3 GIS technology components and complementary areas of interest for the institutional 'triple helix' of the Greek GIS community

Components of GIS technology/GIS community	*Government and utilities*	*Academia*	*Private sector*
Hardware			X
Software		X	X
Digital data	X		X
Skills		X	X

for both central and local government organisations. As a result of this kind of consultancy work teams based in university laboratories have also been active in the effective customisation of commercial GIS software complementing the role of GIS software vendors and engineering consulting firms.

Finally Table 5.3 shows that GIS vendors and engineering consulting firms are primarily interested in selling hardware and software, as well as providing specialised services related to geographic information handling and analysis. Like the academics they have also played a part in raising awareness about GIS at different settings throughout the country, with GIS events like users' conferences (for example, the annual Greek ESRI Users' Conference has been organised by Marathon Data Systems since 1991). The consultants also provide various GIS services in Greece which in some cases demand the development of specific GIS applications including the production of various digital data sets (see, for example, Geomatics www.geomatics.gr) which complement those provided by central government agencies. Moreover a few engineering consulting firms have organised various GIS training seminars, complementing various teams of academics in the provision of GIS training for government officials (see, for example, Epsilon, www.epsilon.gr/gis.htm).

5.4 What stage of development has the Greek GIS community reached?

So far the discussion has focused on the emergence of the Greek GIS community with particular reference to its main actors/stakeholders. The short history of the Greek GIS community can also be divided into four stages: namely, to use the terminology developed by Rogers (2003: 246), cosmopolites, innovators, early adopters and early majority. Table 5.4 shows the cumulative number of GIS actors in Greece in terms of the four stages in the development of the Greek GIS community. In stage I no commercial GIS software is available but individuals from a broad range of institutions studied or worked abroad. In stage II Marathon Data Systems was set up in 1984 to start marketing the ESRI's GIS software and individuals from government organisations, universities and private firms started organising and participating in various GIS events, such as the first GIS workshop in

Table 5.4 Four stages in the emergence of the Greek GIS community and cumulative number of GIS teams

Stage	*Years*	*Cumulative number of GIS teams in Greece*
I. Cosmopolites	From late 1960s to early 1980s	None
II. Innovators	From 1982 to 1985	1
III. Early adopters	From 1986 to 1989	10
IV. Early majority	From 1990 to 1994–5	51

the Geography Lab of the Faculty of Surveying Engineering in the NTUA (Koutsopoulos, 1982). There were nine early GIS adopters in stage III, while in stage IV the cumulative number of GIS adopters significantly increased from 10 to 51.

Stage I: cosmopolites (from late 1960s to early 1980s)

The pre-GIS period (stage I) in the development of the Greek GIS community approximately spans two decades, from the 1960s, when Tomlinson (1988) used the term GIS for the first time to describe the Canadian GIS (see also Coppock and Rhind 1991), to the early 1980s, when GIS adoption started in Greece and other European countries (Masser 1992).

Many individuals who played a central role in the development of the Greek GIS community in the later periods were abroad during this period. Academics, for example in the School of Surveying Engineering at the NTUA and Department of Topography at the AUT, and vendors were cosmopolites who socialised abroad during stage I. As a result they were exposed to new scientific theories and methods, and gained significant GIS-related knowledge and expertise, as well as developing personal contacts with key members of the international GIS community. A good example is Professor Nikos Polydorides, head of the Laboratory of Spatial Planning at the Faculty of Engineering, University of Patras, who carried out his master's dissertation at the Graduate School of Design at Harvard University in 1969–70. At that time Jack Dangermond, the founder and president of ESRI in California, studied for his master's in the same department under Professor Carl Steinitz (1993). The 'weak' non-overlapping network ties (Granovetter 1973, 1982) of these individual 'cosmopolites' have played a significant role in the subsequent development of the Greek GIS community (Assimakopoulos 1997b).

Stage II: innovators (from 1982 to 1985)

Rogers (2003: 248) defines innovators as a very small minority (only 2.5 per cent of the total number of adopters of a particular technological innovation) who are entrepreneurial. Without them new ideas and technologies would never spread. Innovators usually have friendships and contacts outside their local social circles and over great distances. As a result they can cope with the uncertainty of adopting a new idea when none or very few others have adopted it locally. In Greece commercial or proprietary GIS software was not available until 1984.

Stage II of the development of the Greek GIS community is characterised by a period of fermentation for commercial GIS software adoption. Marathon Data Systems was established in 1982 and became the Greek ESRI vendor in 1984. At the same time a professor of geography at the NTUA (Koutsopoulos 1982) published the first GIS paper in the Greek language, while Jack Dangermond, president of ESRI in California, visited the Hellenic Military Geographical Service (HMGS) and made the surveying engineers and government officials there aware of the potential of GIS for both military and civilian cartographic applications. Dangermond's visit and face-to-face discussions with government officials in 1982 at the national mapping agency, coupled with the creation of Marathon Data Systems in 1984 as the Greek ESRI vendor, led to the installation of the first GIS software licence at the HMGS in 1986.

During stage II, a small number of academic laboratories such as the Laboratory of Spatial Planning at the University of Patras used computers and a combination of CAD and relational database management systems software to handle and analyse geographic information and carry out research for automated mapping and urban and regional planning. One of the wider contextual trends which facilitated this development was the first phase of the EPA programme (a major public planning initiative which had as its objective to produce extension plans for the 200 larger Greek cities) organised by the Ministry of Environment, Planning and Public Works. Some central government agencies such as the Hellenic Navy Hydrographic Service also started developing their own software for automated mapping and spatial analysis during this period.

Stage III: early adopters (from 1986 to 1989)

Early adopters represent 13.5 per cent of the total number of adopters according to Rogers' diffusion model. At the system level they are the most significant group of adopters because the necessary critical mass to sustain the diffusion of an innovation would never be reached without them. Rogers (2003: 248) also argues that early adopters are a more integrated part of the local social system than innovators. They are not necessarily cosmopolites but they usually occupy central positions in the structure of the social system within which the innovation diffuses. They are respected by their peers because they often convey to them subjective evaluations of the innovation, playing the role of opinion leaders for potential adopters. The third period of the development of the Greek GIS community is called the period of early adopters because a small group of nine teams from central government agencies, university laboratories and private sector firms started implementing various GIS applications for topographic base-map production and urban planning while there was very little or often no GIS experience at all in similar organisations throughout the country.

More specifically, the Digital Cartography Department at the Hellenic Military Geographical Service first adopted GIS software in 1986. The same year Kalamata City Council in the south of Peloponnese also invested in GIS software after a devastating earthquake as part of its reconstruction programme with EC funding. A group of academics based at the Laboratory of Spatial Planning at the University of Patras adopted GIS software in 1987 for education and research purposes, as a natural progression in their interest in computers in planning (Polydorides 1993). The Hellenic Navy Hydrographic Service of the Ministry of Defence invested in GIS in 1988 following the example of HMGS. In the same year the Committee of Byzantine Antiquities, in cooperation with Rodhos (Rhodes) City Council and the United Nations Environmental Programme, started developing a GIS for the protection of the medieval part of the city of Rodhos.

Two other central government agencies, the National Mapping and Cadastre Organisation (www.okxe.gr) and the National Institute of Mineral and Geological Research (www.igme.gr) installed GIS software and hardware in 1989, taking advantage of EC funding (Mediterranean Integrated Programme) while Nikea City Council in Athens also adopted GIS, taking advantage of the EC Social Fund. Finally, the first private sector engineering consulting firm Eratosthenis successfully used GIS for the production of the extension plan of Patras City Council under the EPA programme. The vast majority of early adopters (eight out of nine teams) adopted ESRI's ArcInfo software. Only the team at Rodhos adopted Mapgrafix for its Apple computers. As a result, ESRI's ArcInfo software was installed in most central government agencies, municipalities, utilities, universities and private sector firms during the next stage (IV) of GIS diffusion in Greece.

Stage IV: early majority (from 1990)

Finally, according to Rogers' model, the early majority represent 34 per cent of the total number of adopters. The early majority interacts frequently with their peers but seldom hold leadership positions (Rogers 2003: 249). Most members of this group often want to be seen neither as the last to lay the old aside, nor as the first to try the new. Stage IV of the development of the Greek GIS community which began in 1990 and continued throughout the 1990s is called the period of early majority for two interrelated reasons: first, because the diffusion of GIS software significantly speeded up and, second, because many more people and institutions, from various disciplinary backgrounds and from all over the country, started interacting by exchanging information, ideas and knowledge about GIS adoption and implementation. The pattern of GIS adoption and implementation in the period of early majority became much more complicated as approximately 40 new teams were added to the list of organisations that formed the Greek GIS community up to the mid-1990s.

These heterogeneous stakeholders constituted part of the early majority of GIS adopters in Greece. In the early 2000s they might, together with the innovators and early adopters, add up to the critical mass that is necessary to sustain further GIS diffusion in Greece. According to Rogers (1991: 253) the critical mass for an interactive computer-based technology like GIS is reached when approximately 20 per cent of the total number of adopters have adopted the innovation. Apart from MDS, which organised its first Greek ESRI users' conference in 1991, other GIS software vendors (e.g. Intergraph, Laserscan, MapInfo) came into the GIS market during this period and many more government organisations also invested in GIS. From central government the Directorates of Forestry under the Ministry of Agriculture, Environment and Urban Planning under the Ministry of Environment, Planning and Public Works as well as the Hellenic Telecommunications Organisation developed various GIS applications.

An increasing number of municipalities (for example, Patras, Volos, Kos, Piraeus, Lamia and Thiva) also adopted and implemented GIS during this period and various teams of academics based at university laboratories in Thessaloniki, Patras and Heraclion in Crete were sub-contracted to develop GIS applications for municipalities or other government agencies. Last, an increasing number of private sector engineering consulting firms such as Eratosthenis, Epsilon, Geomatics, Gaia and Terra began developing GIS applications for central and local government authorities as well as the private sector.

Evaluation of the dynamics of the Greek GIS community

Figure 5.1 shows the cumulative number of GIS actors in Greece from 1982 to 1995. It indicates that the development of the Greek GIS community follows the 'S' shaped logistic curve proposed by Rogers (2003) in his theoretical model of the diffusion of innovations, and by Crane (1972) in her work on the growth of scientific communities and their invisible colleges. The emergence of the Greek GIS community and the formation of its critical mass is illustrated in Figure 5.2(a), (b) and (c). These show the innovators (stage II), early adopters (stage III) and early majority (stage IV) using as a criterion the year in which members adopted commercial GIS software. Apart from the time of GIS software adoption, Figure 5.2 also shows the following heterogeneous dimensions: institutional setting, geographical location, type of GIS software and disciplinary background of the teams shaping this new GIS technological community and tradition of practice.

As can be seen from Figure 5.2(a) in stage II, only Marathon Data Systems adopted commercial GIS software in Greece. Then in stage III, eight teams adopted GIS software, most of them ESRI's ArcInfo. The majority of these

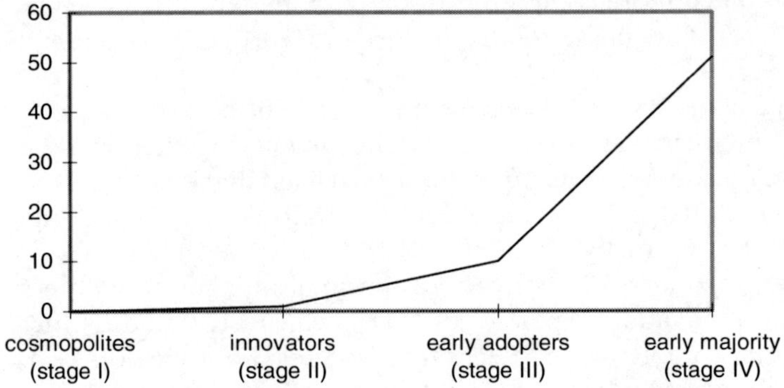

Figure 5.1 The cumulative number of GIS teams according to the four-stage model of the Greek GIS community from 1982 to 1994–5.

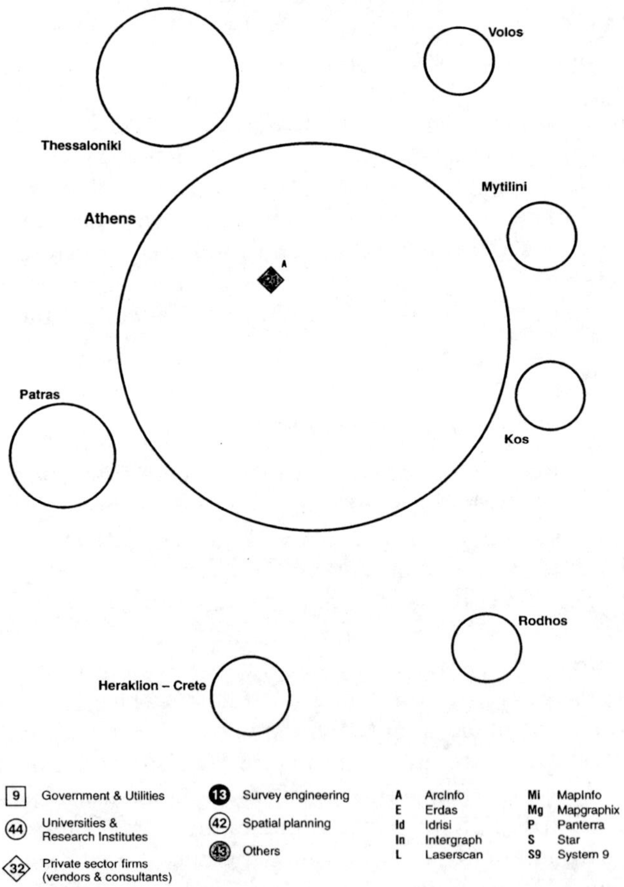

Figure 5.2a The emergence of the Greek GIS community with respect to GIS software adoption in stage II, innovators, 1982–5.

early adopters came from Athens, were based in government institutions and shared a surveying engineering background (see Figure 5.2(b)). Finally, in stage IV, 41 teams adopted GIS software, forming part of the early majority of the Greek GIS community. As can be seen from Figure 5.2(c) these teams came not only from Athens but also from six other cities throughout Greece. From the early 1990s the pattern of GIS diffusion and adoption in Greece becomes increasingly complex as actors from a broad range of institutional settings and an increasing number of locations and disciplines joined the Greek GIS community.

As in Crane's model of invisible colleges the teams which formed the emerging Greek GIS community in the early 1990s came from various geographical locations throughout the country and maintained contacts with

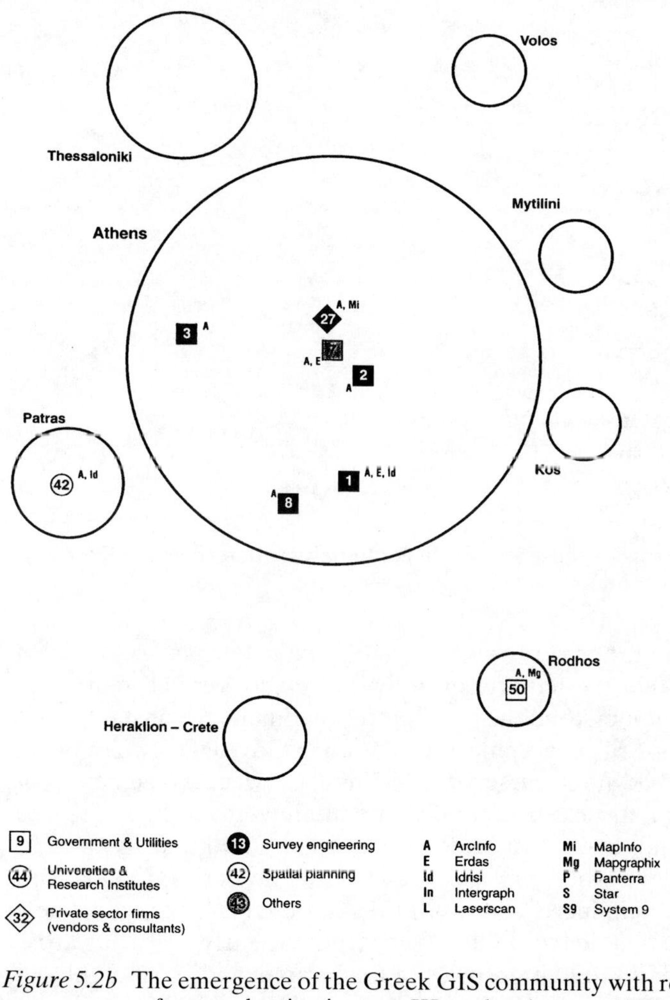

Figure 5.2b The emergence of the Greek GIS community with respect to GIS software adoption in stage III, early adopters, 1986–9.

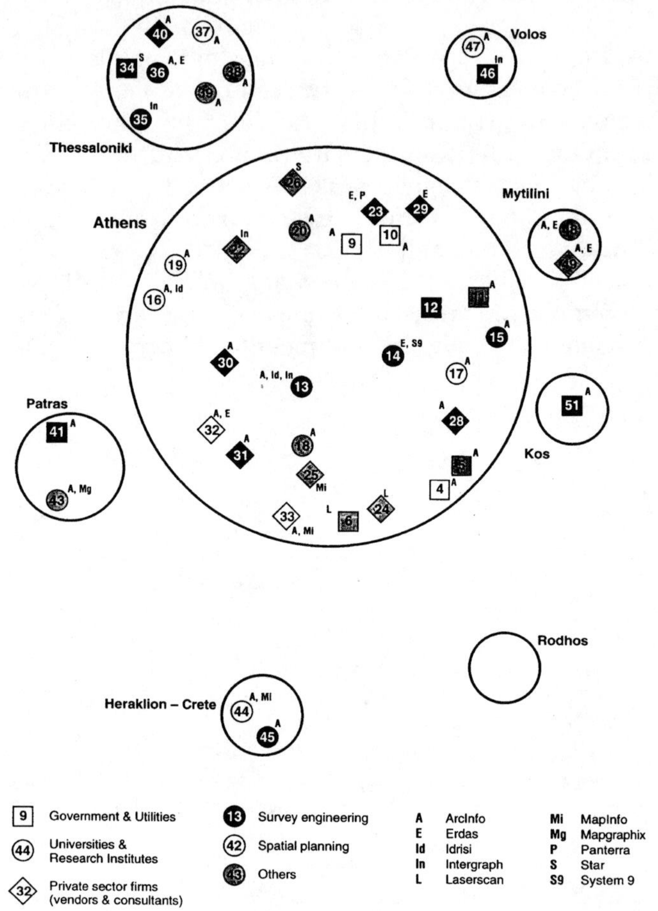

Figure 5.2c The emergence of the Greek GIS community with respect to GIS software adoption in stage IV, early majority, 1990–4/5.

people and institutions abroad over great geographical distances. It is also worth highlighting that the handful of individuals who were cosmopolites and innovators first understood the presumptive anomalies (Constant 1980) related to the handling of geographic information in Greece for two reasons: first, because they had inter-personal GIS linkages with key members of the North American and European GIS communities from the late 1960s onwards; and, second, because a large part of this small minority of individuals were academics who were very close to the theoretical foundations of their respective disciplines. As a result they initiated and fostered a new tradition of practice related to GIS in Greece from the early 1980s onwards.

Academics based in the Laboratories of Geography and Cartography at the NTUA, such as Professors K. Koutsopoulos, G. Veis, and M. Kavouras, have had a strong interest in GIS and digital cartography since the 1970s and early 1980s. In addition some other academics outside Athens such as Professors V. Livieratos at the Department of Cadastre, Photogrammetry and Cartography at AUT and N. Polydorides at the Laboratory of Spatial Planning at University of Patras were some of the people who played a key role in GIS diffusion in Greece as they organised seminars and GIS events from the late 1980s onwards. They also implemented GIS innovations in a number of settings, such as municipalities. Government officials, such as Dr G. Halaris, who is the database administrator of the Digital Cartography Department at the Hellenic Military Geographical Service, adopted ESRI's ArcInfo as early as 1986. Finally, private sector vendors and consultants, such as Mr. A. Kontos founder of Marathon Data Systems and Dr T. Doganis founder of Terra, were among the first individuals to become committed to the idea of fostering a new GIS tradition of practice in Greece, starting up firms and teams who adhered to this new 'revolutionary' technological tradition of practice.

Although these cosmopolites and innovators of the Greek GIS community came from different institutional and disciplinary backgrounds, they had one feature in common. Most of them, like the new 'Argonauts' (Saxenian 2006), have maintained their contacts and work relationships with key members of the international GIS community since the late 1960s. These have operated over great geographical distances transcending national boundaries and promoting the exchange of information and other resources about GIS adoption and implementation between the centres of production in North America and Europe, and Greece. They have also cross-fertilised different traditions of practice related to GIS technology in Greece linking separate individuals, groups or organisations.

In contrast to Crane's (1972) model of invisible colleges in science, where new communities of scientists tend to be rather homogeneous adhering to a particular paradigm (Kuhn 1970, Gutting 1984), there is a great deal of heterogeneity in the construction of the new Greek GIS community, since academics, government officials and private sector vendors and consultants from various disciplinary backgrounds have participated in its development from the early 1980s onwards, maintaining multiple memberships in a range of pre-existing technological communities as well as the emerging new GIS community. Issues of both heterogeneity (Law 1994) and complementarity are explored next in greater depth, based on the patterns of GIS linkages among the different GIS teams and institutional and disciplinary groups forming the Greek GIS community.

5.5 Heterogeneous linkages within the Greek GIS community

Figure 5.3 shows the 95 linkages that connect the 51 teams who form the Greek GIS community, highlighting the essential complementarities of interests among different teams. These linkages represent about 7.5 per cent

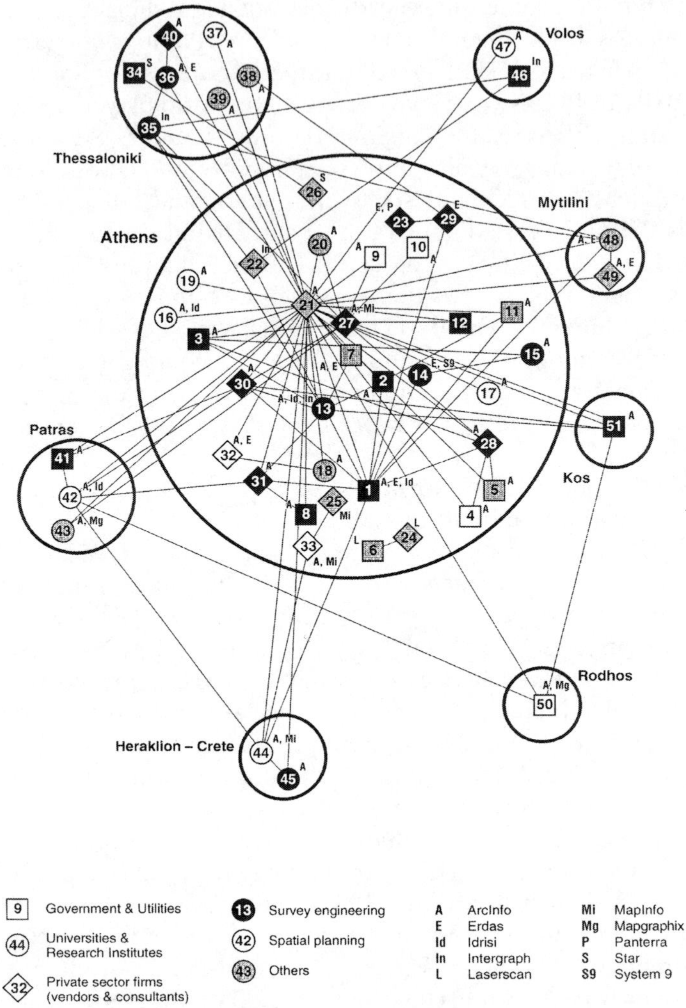

Figure 5.3 GIS linkages of the GIS teams forming the Greek GIS community.

of the maximum number of linkages between the members of the Greek GIS community and usually have a multi-stranded nature, as GIS technology consists of various components such as hardware, software and digital data. They are grouped according to economic, knowledge and social criteria in three types:

- contractual relations of a commercial nature (i.e. supplier–customer, contractor–customer) with respect to the delivery of GIS hardware, software, digital data and services;

- knowledge relations such as between teachers and students with respect to the creation and sharing of GIS-related knowledge skills and expertise; and
- social relations of awareness of other people's work and ideas, friendship and support in terms of GIS adoption and implementation.

These heterogeneous socio-economic and knowledge linkages were derived from the materials collected as a result of in-depth semi-structured interviews with 70 GIS experts, team leaders and experts throughout Greece, who were interviewed at least once (some more than twice over a period of two to three years from summer 1992 to summer 1994). These enabled the identification of the commercial relations with respect to various GIS components such as hardware and GIS software, digital data and services for the development of GIS applications throughout Greece. Knowledge relations, such as teacher–student relations, were also identified from these interviews which asked about academic qualifications and attendance of short courses and seminars organised by various university-based teams throughout the country, as well as unobtrusive methods such as co-authorship data. Finally, information about social relations, such as friendship and awareness of others' work and ideas, emerged not only from the semi-structured interviews but also from participant observation data from various GIS events, such as seminars, short courses and users' conferences.

However, it should be noted that Figure 5.3 does not depict the multiple contents of GIS linkages between the 51 teams. As Wasserman and Faust (1994) and others have pointed out, this is a common limitation of network research as the untangling of the different strands in multiplex relationships which convey not only information but also material resources and services is usually very difficult. Figure 5.3 also shows relations with no direction, tacitly assuming that all 95 linkages are reciprocal in nature. Mutuality or symmetry of linkages is not always the case between the various teams in terms of GIS adoption and implementation. Power relations are usually built because of the asymmetry of linkages with respect to the exchange of information and other resources (Callon 1993; Law 1991). However, Figure 5.3 does not consider the bidirectional nature of linkages because of the practical difficulties involved in estimating the intensity of such linkages. There is also no discussion of time with respect to the evolution of the patterns of linkages in Figure 5.3. This reflects the difficulties experienced by interviewees in studies such as this of recalling with any precision how they developed their linkages over time (Bernard *et al.* 1984; Doreian 1995). However, despite these limitations it is felt that Figure 5.3 gives a good indication of the overall structure of the critical mass of the Greek GIS community in the mid-1990s.

Figure 5.4 explores the distribution of linkages in Figure 5.3. It shows that the total number of internal linkages is 190 as it is assumed that all 95 linkages connecting the 51 teams at Figure 5.3 are reciprocal in nature. Figure 5.4 also

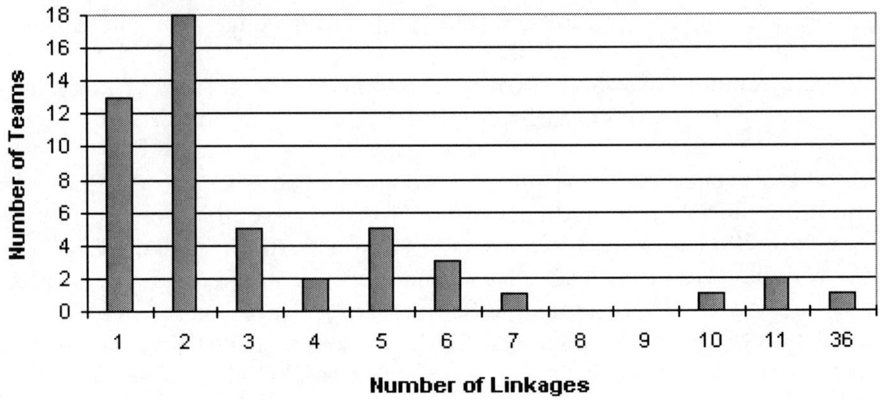

Figure 5.4 The distribution of linkages within the Greek GIS community.

shows that the majority of teams have only one or two linkages, whereas four teams account for over a third of all linkages. These four teams are: 1, the Digital Cartography Department at the Hellenic Military Geographical Service; 13, the Laboratory of Cartography at the NTUA; 21, Marathon Data Systems; and 27, the engineering firm Eratosthenis. Marathon Data Systems (team 21) alone accounts for nearly a fifth (19 per cent) of the total number of linkages (or 36 out of 190 internal linkages). Because these four teams seem to play a very central role in providing connectedness to the whole of the Greek GIS community the centrality and prominence of teams within that community is explored next with socio-metric techniques and SNA software.

5.6 Prominence and centrality

The findings of the previous analysis show that teams such as Marathon Data Systems, the Greek ESRI vendor, occupy central positions in the structure of the Greek GIS community. To explore in more detail the prominence and centrality of these actors and relevant groups the following questions are considered below:

- Is it possible to understand the prominence of teams and relevant social groups from an analysis of cliques in the Greek GIS community?
- How central are the teams who come from the relevant social groups shaping GIS technology in Greece?

Prominence

As discussed in the Appendix, cliques can be defined in a social system according to the complete mutuality of linkages of the subgroup members.

A clique consists of at least three mutually connected actors. In Figure 5.3 Ucinet 6 (Borgatti, Everett and Freeman 2002) was used to identify cliques of a minimum of three teams. As a result Figure 5.5 shows a list of 31 cliques of the Greek GIS community and its member teams. As can also be seen from Figure 5.5, eight out of these 31 cliques consist of four teams; hence there are also eight cliques of four within this GIS community. Membership in a large number of cliques indicates prominence, as members of teams who share many cliques have more opportunities to exchange information and other resources with respect to GIS adoption and implementation.

Figure 5.6 shows a cohesive subgroup of the Greek GIS community based on the ranking order of teams who participate in more than 10 per cent of the total number of cliques of three. In a sense any classification of cohesive subgroups is arbitrary and this is no exception. The criterion for being a member of a cohesive subgroup could be set higher than 10 per cent of the total number of cliques of three but the aim here is to include all teams who share a considerable number of cliques so that the prominence of particular teams and groups related to GIS technology in Greece can be explored.

From this it can be seen that the four teams (1, 13, 21 and 27) with the highest number of internal linkages share a large number of cliques of three and as a result feature prominently in Figure 5.5. Marathon Data Systems (team 21) is a member of 90 per cent (or 28 out of the 31) of the cliques of three. Teams 1 at the Digital Cartography Department of the Hellenic Military Geographical Service, 27 at the engineering consulting firm Eratosthenis, and 13 at the Laboratory of Cartography at the NTUA, are respectively members in 26, 23 and 19 per cent of cliques of three. Additionally teams 28 and 30, at Geomatics and Terra engineering consulting firms, and 42, at the Laboratory of Spatial Planning at the University of Patras, are each a member of 13 per cent (or four out of 31) of cliques of three.

```
 1:  1 2 13 21           17:  20 21 27
 2:  1 13 21 30          18:  21 27 30 51
 3:  1 13 21 31          19:  21 27 43
 4:  1 2 11 21           20:  21 36 40
 5:  1 21 28             21:  21 30 41 42
 6:  1 21 44             22:  21 42 50
 7:  1 21 48             23:  21 31 42
 8:  3 21 27 28          24:  21 42 44
 9:  3 13 21             25:  21 44 45
10:  4 21 28             26:  21 48 49
11:  5 21 28             27:  21 50 51
12:  7 13 21             28:  13 21 30 51
13:  8 21 31             29:  13 22 35
14.  9 21 27             30:  22 35 46
15:  12 21 27            31:  1 29 48
16:  18 21 32
```

Figure 5.5 The list of 31 cliques of the Greek GIS community.

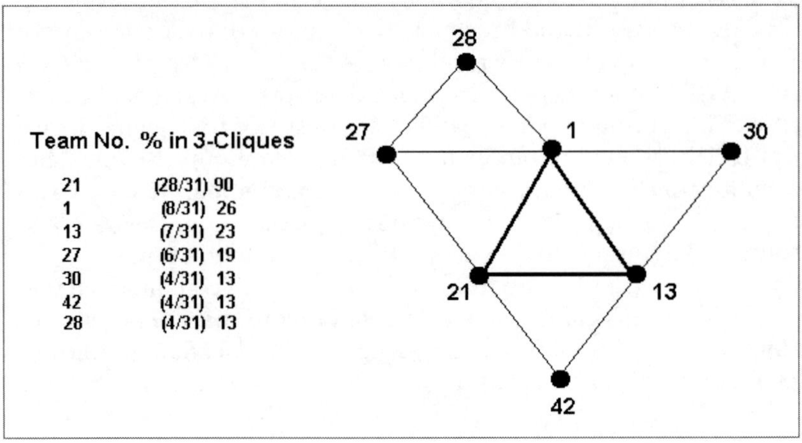

Figure 5.6 Cohesive subgroup of the Greek GIS community based on the ranking of teams who are members in more than 10 per cent of cliques of three.

The heterogeneity of this cohesive subgroup can be explored along the key dimensions mentioned above (i.e. institutional and disciplinary backgrounds, GIS software adopted, geographical location/city). In terms of institutional setting the majority of teams (four out of seven) in Figure 5.6 are GIS software vendors and consultants. Then there are the two academic teams at the NTUA and the University of Patras, and the government team at the national mapping agency. This highlights once more that a handful of teams, mainly from the private sector and academia, transcended the institutional boundaries of the 'triple helix', participating in a number of cliques of three in the Greek GIS community. This undoubtedly gives opportunities for members of these teams to share and exchange information about GIS technology. The vast majority of prominent teams (six out of seven) in Figure 5.6 are based in Athens and all seven teams have adopted ESRI's ArcInfo GIS software. Finally, teams that share a survey engineering background are the majority (five out seven) in Figure 5.6, with only one team (42) from a spatial planning disciplinary background and one team (21) with one of the other backgrounds related to GIS technology in Greece.

Centrality

As also pointed out in the Appendix, the centrality of actors in a graph can be measured in various ways such as degree, closeness, betweenness and flow betweenness. Figure 5.7 shows the findings for these centrality measures for the 51 teams. The degree centrality measure shows the number of linkages that each team has in Figure 5.3. From this it can be seen that the four

Team No.	Degree	Closeness	Betweenness	Flow Betweenness
1	11	17.36	9.04	8.20
2	4	16.72	0.08	1.08
3	5	16.78	3.19	4.01
4	2	16.29	0.00	0.29
5	2	16.29	0.00	0.29
6	1	2.00	0.00	0.00
7	2	16.45	0.00	0.15
8	2	16.29	0.00	0.53
9	2	16.45	0.00	0.15
10	1	14.84	0.00	0.00
11	3	16.50	0.00	0.68
12	2	16.45	0.00	0.15
13	10	17.18	6.06	5.69
14	2	15.15	0.52	2.28
15	2	14.79	0.12	2.28
16	1	16.23	0.00	0.00
17	1	16.23	0.00	0.00
18	2	16.29	0.00	1.70
19	1	16.23	0.00	0.00
20	2	16.45	0.00	0.15
21	36	19.01	61.80	48.18
22	3	15.20	0.70	2.40
23	1	13.51	0.00	0.00
24	1	2.00	0.00	0.00
25	1	12.95	0.00	0.00
26	1	2.00	0.00	0.00
27	11	17.12	6.95	10.69
28	6	16.89	0.88	4.12
29	4	15.38	3.83	4.50
30	7	16.95	0.93	1.94
31	5	16.78	0.43	2.23
32	2	16.29	0.00	1.70
33	2	14.66	3.67	2.14
34	1	2.00	0.00	0.00
35	6	15.82	3.96	6.04
36	3	16.56	0.85	2.25
37	1	16.23	0.00	0.00
38	2	16.45	1.25	0.92
39	1	16.23	0.00	0.00
40	2	16.39	0.00	0.92
41	0	16.04	0.00	0.49
42	6	16.61	0.58	3.03
43	2	16.45	0.00	0.15
44	5	16.78	7.41	6.40
45	2	16.39	0.00	0.63
46	2	13.89	0.00	0.92
47	1	16.23	0.00	0.00
48	5	16.89	3.10	3.17
49	2	16.50	0.00	0.50
50	3	16.34	0.03	0.83
51	5	16.67	0.28	1.34

Figure 5.7 Centrality measures (degree, closeness, betweenness and flow betweenness) for the 51 teams who form the Greek GIS community (calculated by UCINET 6, Borgatti, Everett and Freeman 2002).

teams with a high number of internal linkages (21, Marathon Data Systems; 27, the engineering consulting firm Eratosthenis; 1, Digital Cartography Department at the Hellenic Military Geographical Service; and 13, Laboratory of Cartography at the NTUA) are the clear leaders and that most of the other teams have a low degree centrality.

Unlike degree centrality, the closeness centrality scores in Figure 5.7 show that the large majority of teams are quite close to the four leading teams (21, 27, 1 and 13) even when they have only one internal linkage. For example, the closeness centrality score of MDS, the Greek ESRI vendor (team 21), is 19.01, while the score of team 16 at the Laboratory of Geography and Regional Planning at the NTUA, which has only one linkage with team 21, is 16.23. A tentative explanation for this finding is that the large majority of teams who form the Greek GIS community have adopted the same GIS software (ArcInfo). Therefore they know each other from various informal and formal GIS events like the Greek ESRI Users' Conferences. As a result the communication paths between the large majority of teams are very short (one or two linkages) and hence their closeness centrality is quite high.

Even some of the teams that have adopted different GIS software, like team 35 at the Laboratory of Cartography at the AUT, have a relatively high closeness centrality (15.82) because they maintain direct linkages with the leading ArcInfo teams. The only notable exceptions in Figure 5.7 with a very low closeness centrality (2.00) are the two dyads of customer–supplier teams (6, Directorate of Forests at the Ministry of Agriculture, and 24, Space Hellas – the vendor of Laserscan's GIS software; 34, Technical Services of Thessaloniki City Council, and 26, the vendor of Star's GIS software) that are not linked with the Greek ESRI, or one of the other teams with a high number of internal linkages, and were characterised as isolates in Figure 5.3.

Finally in terms of betweenness, and more importantly, flow betweenness centrality Figure 5.7 shows that a small minority of teams (21, 27, 1 and 44 at the Regional Planning Division of the FoRTH) are clear leaders who broker relationships and become bridges between many teams of the Greek GIS community. It is worth noting here that betweenness centrality considers how often an actor lies in the shortest paths connecting all pairs of other actors in a network. Flow betweenness, however, considers how often an actor lies not only in the shortest paths, but all paths. As a result flow betweenness seems a better measure than betweenness for teams of the Greek GIS community because there is no reason to believe that individual members of teams restrict their communication with respect to GIS technology to the shortest path between each pair of teams.

As can be seen from the last column of Figure 5.7, three (21, 27 and 1) out of the four teams with the highest number of linkages feature prominently in the communication paths of the other teams controlling the flow of information about GIS innovations in Greece. A surprising finding is that team 44 has higher betweenness and flow betweenness scores than the other teams with the highest number of linkages, but this is perhaps justified on the ground that the head of team 44 is not only an academic, but also a vendor, who established Omas the Greek MapInfo vendor, and a consultant who owns the engineering consulting firm InfoCharta. On the other hand most of teams have zero (or very low) betweenness and flow betweenness centrality. This is because the large majority of teams have a low number of internal linkages and rarely stand between other teams. For example team 16 at the Laboratory of Geography and Regional Planning at the NTUA has only one internal linkage with team 21 (Marathon Data Systems) and its betweenness and flow betweenness scores are zero.

Figure 5.7 further confirms that team 21 (Marathon Data Systems) is by any measure the most central team of the Greek GIS community. However, this finding generates doubts about how the relationship between centrality and the prominence of specific teams should be interpreted against the structure of the whole system when too much faith is put in the mechanistic interpretation of numbers. Individual actors like the ESRI vendor are not so influential for the development of the whole of the Greek GIS community if the effects of the structure of this community are taken into account. It should be kept in mind that this team neither shares the predominant surveying engineering background nor belongs to academia, which maintains most of the external linkages with GIS stakeholders abroad. As a result it can be argued that its prominence and prestige is disproportionately represented by the scores shown in Figure 5.7.

It might be argued that teams which have lower centrality scores than team 21 but belong to the dominant social group of surveying engineers may be better placed to influence the development of the Greek GIS community intellectually and politically. Consider, for example, the surveying engineering teams based at university laboratories such as team 13, the Laboratory of Cartography at the NTUA, or at government agencies, such as team 1, the Digital Cartography Department at the HMGS, or at engineering consulting firms, such as team 27, Eratosthenis. These three teams (1, 13, 27) occupy central positions in the inner circle of the Greek GIS community, as will be discussed below, belong to most of its cohesive subgroups (see Figure 5.5) and feature prominently in Figure 5.6. Because they share with the majority of teams the same disciplinary background and the technological tradition of practice related to GIS, they may be in a better position than team 21 to influence the overall development of the Greek GIS community intellectually, politically and technically. This can be seen in the setting up of such influential institutions as the National Mapping and Cadastre Organisation which was set up by surveying engineers from the NTUA and the Ministry of Environment and Planning.

From the outset, therefore, it seems useful to make a distinction between 'monomorphic' opinion leaders (Rogers 2003: 288) who can merely influence the adoption of GIS software (e.g. Greek ESRI vendor) and other polymorphic GIS opinion leaders like the academic teams at the NTUA (team 13, Laboratory of Cartography), Thessaloniki (team 35, Laboratory of Cartography) and Patras (team 42, Laboratory of Spatial Planning) who make a significant contribution in many different areas, including the design of new institutions related to geographic information handling and analysis, the implementation of GIS applications in a variety of institutional settings, and the research and teaching of GIS courses.

More importantly the findings of this research support the argument by Rogers and Kincaid (1981: 124) that opinion leadership is more a structural than an individual characteristic. To gain insights into why actors are considered GIS opinion leaders it would be necessary to analyse the patterns of direct and indirect GIS linkages from which opinion leaders elicit their power to influence the adoption and implementation processes of GIS at a local, national and international scale. Nevertheless it is felt that one of the significant contributions of this research is that it shifts the emphasis of the debate about the roles of individuals in GIS innovations from individual 'trees' such as GIS champions (Azad 1997) to the 'forest' of GIS relationships.

5.7 Institutional setting

As pointed out in Section 5.2, the two largest components of the Greek GIS community are the academic and private sector groups of teams. Graph 5.1 shows the positioning of the 51 teams according to the Euclidian distances/ structural equivalence (see Appendix) between them. It also shows the direct connections between these 51 teams. The boxes represent teams from central and local government and utilities organisations. The circles represent teams from academia, and the diamonds represent teams from private

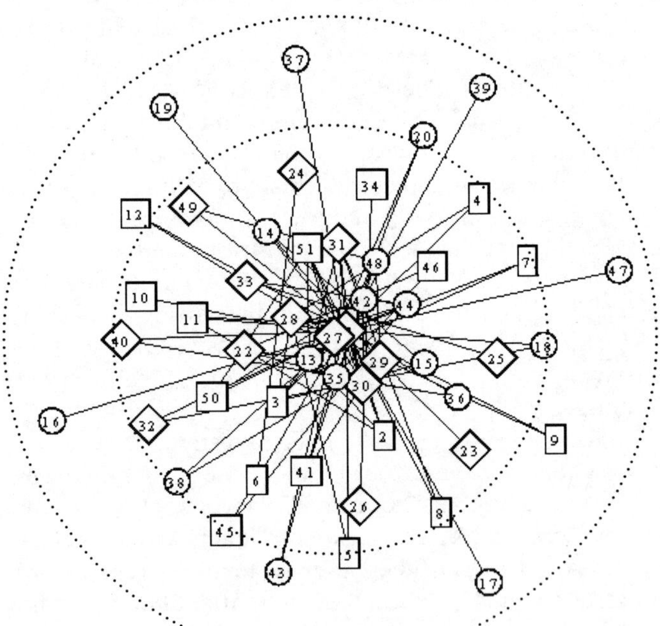

Graph 5.1 Institutional groups of the 51 teams which form the Greek GIS community based on the Euclidian distances between them.

sector companies including GIS software vendors and consultants. Teams that are closer together in Graph 5.1 have similar patterns of connections to the other teams that form the Greek GIS community, and therefore are structurally equivalent. The distance between them is the Euclidian distance (Burt 1987) calculated according to the pattern of internal linkages to and from all the other teams.

Graph 5.2 shows in detail teams which are positioned in the centre of Graph 5.1. The Euclidian distances between them are so small that it is necessary to zoom in on Graph 5.1 to separate them. As can be seen from Graph 5.2, a handful of teams including 21 (Marathon Data Systems), 27 (Eratosthenis) and 1 (Digital Cartography Department at the HMGS) are located near the geometrical centre of Graph 5.1. Marathon Data Systems (team 21) was chosen as the centre of the inner, middle and outer circles of the Greek GIS community for three interrelated reasons: first, it was the earliest adopter of GIS software in Greece; second, it has developed the highest number of direct linkages with other teams of the Greek GIS community; and, third, it is positioned in the centre of Graphs 5.1 and 5.2 as a result of its Euclidian distance from the rest of the teams. Subsequently the radius of the three circles was defined as 1/3, 2/3 and 3/3 of the distance between teams 21 and 16. As can be seen from Graph 5.1, the distance between teams 16 and 21 is the largest Euclidian distance between the teams in the centre and the periphery of the Greek GIS community.

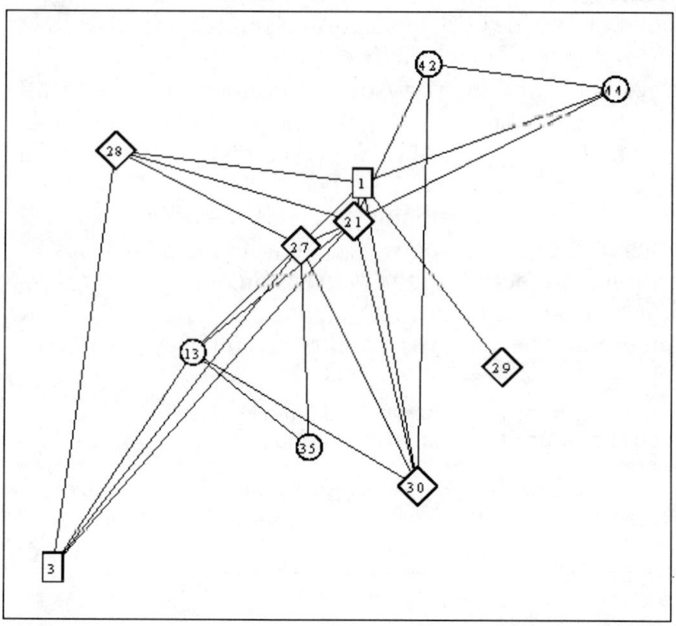

Graph 5.2 Zooming in on the core of the Greek GIS community.

Graph 5.1 shows that there is a great deal of heterogeneity in terms of institutional backgrounds within the inner circle and the whole of the Greek GIS community. Teams of the same shape are not clustered together but form a complicated pattern. It is worth also pointing out that the Euclidian distances in Graph 5.2 between teams from a broad range of institutional settings are small. It seems that members of public and private sector teams as well as academics can easily blur the boundaries between the different sectors of the Greek GIS community as they have similar patterns of connections in collaboration or competition for scarce resources. The 19 teams which belong to the inner circle come from a broad range of institutional settings: 1, 2 and 3 from central government, 46 and 51 from municipalities, 13, 15, 35, 42, 44 and 48 from academia, and 21, 22, 27, 28, 29, 30, 31 and 33 from the private sector. Similar heterogeneity, although to a lesser extent, can be observed for the 23 teams of the middle circle, and the 9 teams of the outer circle.

Table 5.5 shows the percentage of teams in the inner, middle and outer circles of the Greek GIS community according to the main institutional groups. Private sector teams account for 42 per cent of the inner circle because a handful of GIS vendors and engineering consulting firms provide connectedness to the whole of the Greek GIS community. In second position are a handful of teams from academia who account for about a third (32 per cent) of the inner circle. As might be expected, although they form the largest component of the Greek GIS community and have most of its external linkages, academics do not maintain many internal linkages unless it is for the development of specific GIS applications. This is also reflected in the composition of the outer circle in which there are no private firms but 78 per cent come from academia and the rest from government organisations. Government and utilities teams form the largest group in the middle circle accounting for almost half (48 per cent) of it. The rest is divided between the private sector (30 per cent) and academia (22 per cent).

To discuss in greater detail the positioning of the 51 teams in Graph 5.1 according to the Euclidian distances between them and the patterns of internal linkages within and between the various institutional groups, the next two sets of graphs are put forward. Graphs 5.3, 5.4 and 5.5 show the Euclidian distances and the patterns of internal linkages within the

Table 5.5 Percentage of teams in the inner, middle and outer circles of the Greek GIS community according to the main institutional groups

Percentage of teams in the inner, middle, outer circle/institutional groups	*Inner circle (19 teams)*	*Middle circle (23 teams)*	*Outer circle (9 teams)*
Government (18 teams)	26	48	22
Academia (18 teams)	32	22	78
Private sector (15 teams)	42	30	0

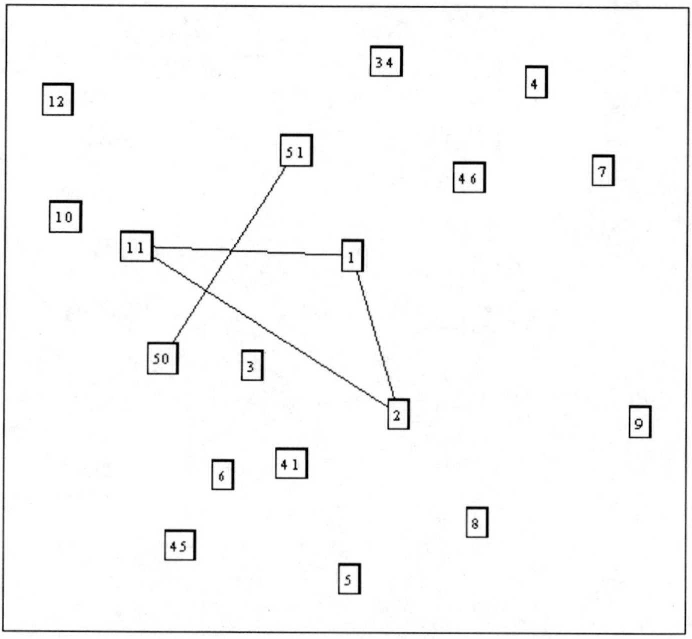

Graph 5.3 Teams which belong in central and local government and utilities organisations based on the Euclidian distances between them.

government, academia and the private sector, respectively. Graphs 5.6, 5.7 and 5.8 show the same information for Euclidian distances and the patterns of internal linkages between these three institutional groups.

Graph 5.3 shows the teams which adopt and implement GIS within central and local government and utilities organisations. As might be expected, this group of teams is to a large extent unconnected. Only the two agencies under the Ministry of Defence (teams 1 and 2 at the Hellenic Military Geographical and Navy Hydrographic Services) maintain linkages between each other and with team 11 (Hellenic Telecommunications Organisation). Given their geographical proximity it is also not surprising to find that teams 50 and 51 within the Technical Services of the municipalities of Rodhos and Kos are linked with respect to GIS adoption and implementation. The rest of the Greek public sector organisations are not connected, although most of them sustain a number of GIS linkages with members of the two other institutional groups, private sector companies and university-based laboratories.

Graph 5.4 shows the positioning of academic teams based on the Euclidian distances between them and the patterns of internal linkages between them. As might be expected, the pattern of GIS linkages between academic teams is more dense than was the case for government teams. About half (8/18) of the academic teams that occupy positions in the inner circle of the Greek GIS community are interconnected. Teams form dyads based on common

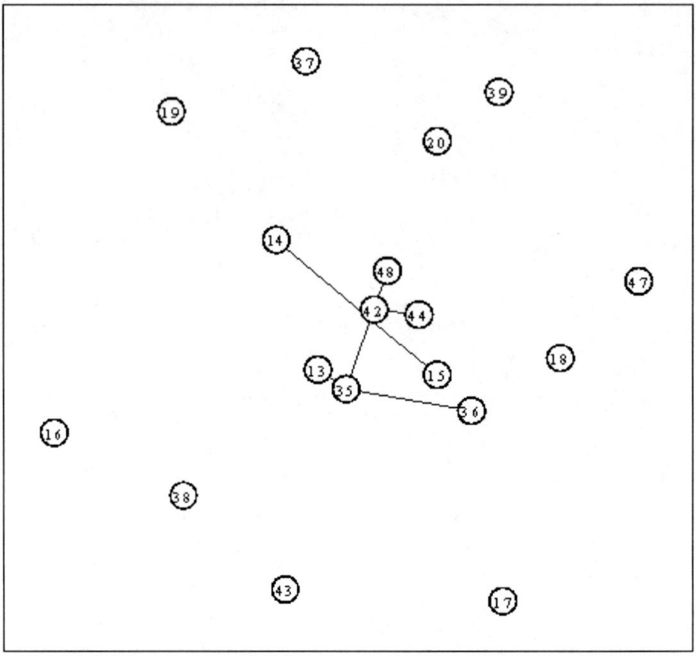

Graph 5.4 Academic teams based on the Euclidian distances between them.

interests about specific types of GIS applications. For example, teams 13 and 35 from the Laboratories of Cartography at the NTUA and Aristotle University of Thessaloniki are linked because they share a common interest in digital cartography and cadastral GIS applications. The same is the case for teams 42 and 44 from the Laboratories of Spatial Planning at the University of Patras and the Regional Planning Division of the FoRTH at Heraclion in Crete with respect to land-use planning.

In contrast to government and academia, Graph 5.5 shows that the majority of teams which belong to the private sector are interconnected. This is mainly due to the activities of the Greek ESRI vendor (team 21, Marathon Data Systems) which is the best linked team from the GIS software vendors with engineering consulting firms. Graph 5.5 shows that this team maintains linkages with teams 27, Eratosthenis; 28, Geomatics; 30, Terra; and 31, Gaia, because of various joint GIS projects throughout the country. In the middle circle of the Greek GIS community there is a small minority of GIS software vendors like Laserscan (team 24) and Star (team 26) which are isolates within the private sector because they have managed to sell their packages to only a very few teams. However the same teams (24 and 26) are not isolates in the next graph which shows the pattern of linkages between the private sector and the government agencies.

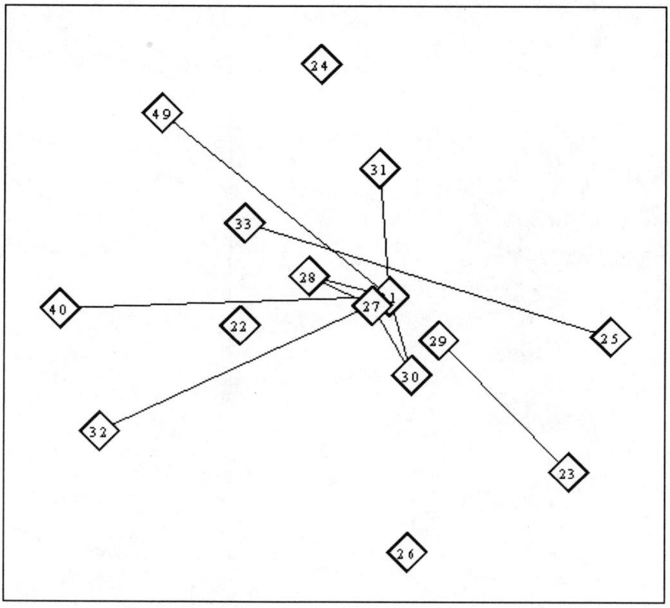

Graph 5.5 Teams which belong in private sector firms based on the Euclidian distances between them.

As pointed out above, the intensity of linkages between government organisations and private sector companies is the highest within the Greek GIS community. As can be seen from Graph 5.6, almost all teams have at least one or two connections. Most of them have a contractual, customer supplier nature, as vendors and consultants supply various components of GIS technology such as hardware and software, and develop various GIS applications for the public sector organisations. In terms of the positioning of teams, the inner circle of Graph 5.6 consists mostly of vendors and consultants. The only exceptions from government organisations are teams 1, 2 and 3 from the Digital Cartography Departments of the Ministries of Defence and Environment, agencies that occupy positions within the inner circle of the Greek GIS community because of their role as digital data producers and suppliers.

Unlike Graph 5.6, more than half of the teams from university laboratories and government organisations shown in Graph 5.7 are not directly linked with respect to GIS adoption and implementation. In fact linkages in Graph 5.7 are concentrated between teams that belong to the inner and middle circles of the Greek GIS community. These linkages mainly reflect either teacher–student relations, or awareness of specific research GIS projects that teams of academics have carried out for government organisations, or collaborative work between academics and central and local government

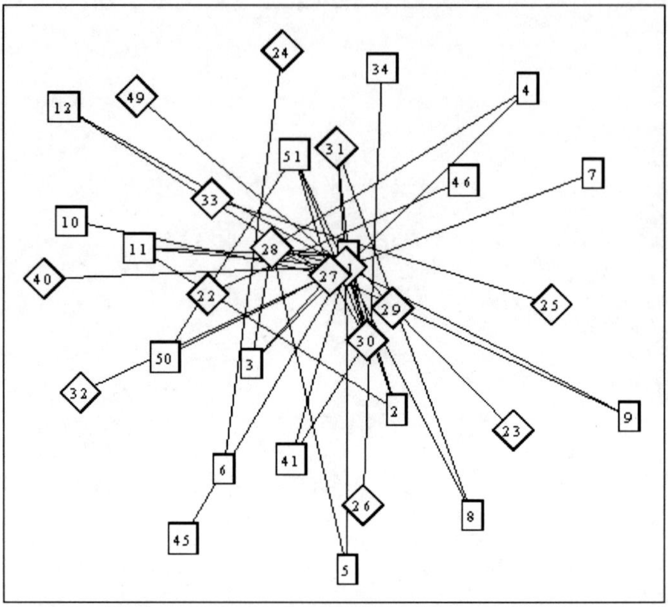

Graph 5.6 Teams which belong in government (boxes) and private sector (diamonds) based on the Euclidian distances between them.

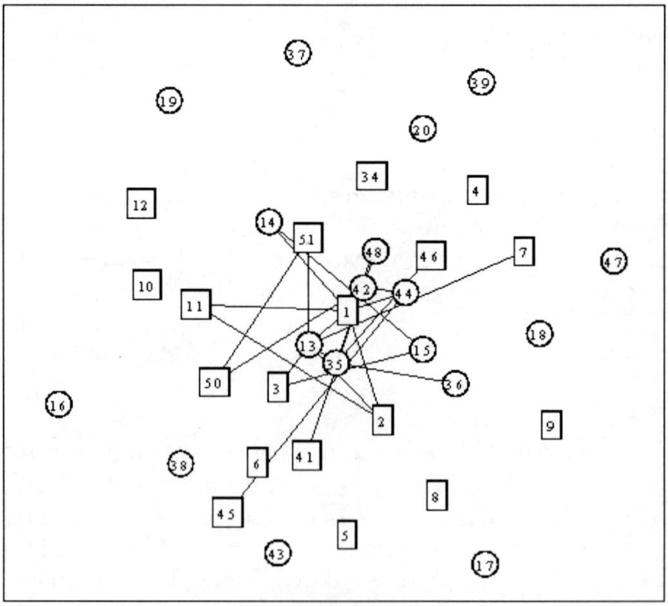

Graph 5.7 Teams which belong in government (boxes) and academia (circles) based on the Euclidian distances between them.

engineers. For example, the linkage between teams 13 (Laboratory of Cartography at the NTUA) and 51 (municipality of Kos) is a teacher–student relation which also has a social component. The head of the municipal GIS Department at Kos wrote her diploma dissertation at the Laboratory of Cartography at the National Technical University of Athens.

Last, the positioning of teams and the patterns of linkages between private sector and academia are worth examining. As can be seen from Graph 5.8, the intensity of GIS linkages between these two social groups stands somewhere between that of the two previous graphs. This can be justified on the grounds that private sector companies understandably maintain the highest number of GIS linkages with the government agencies and vice versa, while university laboratories are not so keen on maintaining GIS linkages with government institutions unless there is collaboration for the development of specific applications. A significant finding from Graph 5.8 is that teams from private firms and university laboratories seem to drive the development process of the Greek GIS community, as 74 per cent of the teams forming the inner circle of this GIS community come from these two social groups (see also Table 5.5).

5.8 Disciplinary background

The group of teams which shares a surveying engineering disciplinary background accounts for 43 per cent of the total number of teams forming the

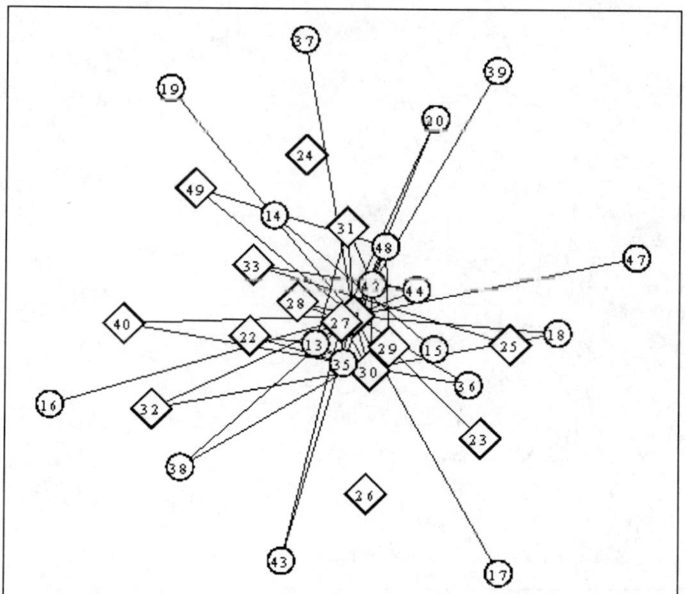

Graph 5.8 Teams which belong in academia (circles) and private sector (diamonds) based on the Euclidian distances between them.

predominant group of the Greek GIS community. Graph 5.9 shows the disciplinary backgrounds of the 51 teams which form the Greek GIS community based on the Euclidian distances between them. The boxes represent teams with a surveying engineering background. The circles represent teams with a spatial planning background based academically and professionally on architecture and civil engineering. And the diamonds represent teams which share such diverse disciplinary backgrounds as human and physical geography, electrical and computer engineering, mathematics, agriculture, forestry and the environment.

Table 5.6 shows the percentage of teams in the inner, middle and outer circles of the Greek GIS community according to these disciplinary groups. From this it can be seen that unlike the findings for institutional setting (see Graph 5.1 and Table 5.5) the inner circle of the Greek GIS community is rather homogeneous with about two-thirds of the 19 teams (or 68 per cent) sharing a surveying engineering background. The other third is equally divided between the spatial planning and the other disciplinary groups. This is a significant finding with considerable implications for this GIS community. It shows that the group of teams with a surveying engineering background is the dominant relevant social group forming the critical mass of the Greek GIS community, because the majority of the most central GIS teams in Greece share a surveying engineering background and maintain a strong

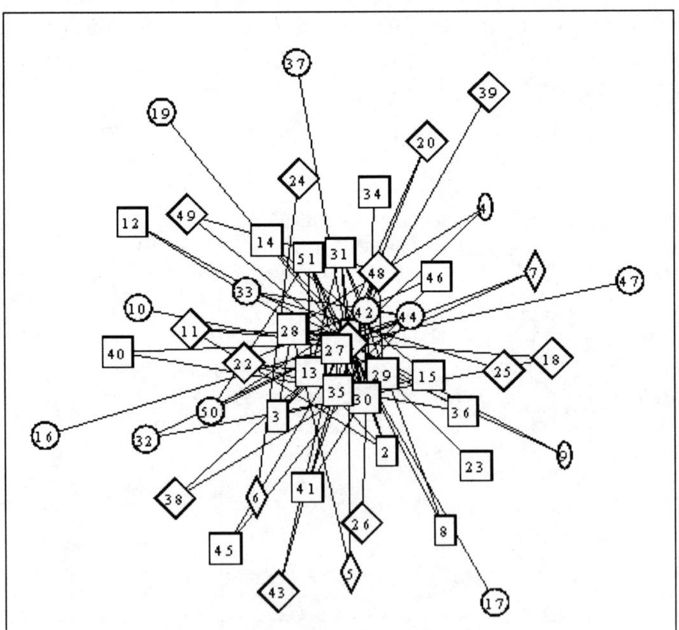

Graph 5.9 Disciplinary backgrounds of the 51 teams which form the Greek GIS community based on the Euclidian distances between them.

Table 5.6 Percentage of teams in the inner, middle and outer circles of the Greek GIS community according to the main disciplinary groups

Percentage of teams in the inner, middle, outer circle/disciplinary groups	*Inner circle (19 teams)*	*Middle circle (23 teams)*	*Outer circle (9 teams)*
Survey engineers (22 teams)	68	35	11
Spatial planners (13 teams)	16	17	67
Others (16 teams)	16	48	22

web of affiliations with the surveying engineering tradition of practice. In the past decade or so, as a result of the centrality of the surveying engineering tradition of practice in the emerging structure of the Greek GIS community, many of the applications and resources have focused unsurprisingly on cadastral and related areas of interest.

As can be also seen from Table 5.6, surveying engineers are the second largest group (35 per cent) in the middle circle, and the third largest (11 per cent) in the outer circle of the Greek GIS community. The spatial planners occupy two-thirds (67 per cent) of the outer circle, while the group of teams with one of the other backgrounds occupies almost half (48 per cent) of the middle circle. The next set of graphs explore in more detail the patterns of direct linkages and the positioning of the 51 teams in Graph 5.9. Graph 5.10 shows the surveying engineering group of teams. Only two out of these 22 teams are unconnected with the rest. The vast majority of teams with a surveying engineering disciplinary background maintain at least one linkage with some other member of the surveying engineering tradition of practice.

As can be also seen from Graph 5.10, the 13 surveying engineering teams which shape a significant part of the inner circle of the Greek GIS community maintain a number of direct linkages between them. It can be argued therefore that this cohesive subgroup of surveying engineering teams has constructed a comparative advantage for surveying engineers in Greece. Its members possess not only hardware, software and digital topographic data but have also produced a shared body of knowledge and expertise related to GIS technology which fuels the further diffusion of GIS throughout Greece according to the surveying engineering tradition of practice. On the other hand, as can be seen from Graph 5.11, more than two-thirds of the teams that share a spatial planning disciplinary background are unconnected with each other and occupy positions in the outer circle of the Greek GIS community.

Only two teams at the Laboratory of Spatial Planning at the University of Patras (team 42) and the Regional Planning Division at the Foundation for Research and Technology Hellas at Heraclion in Crete (team 44) are proximate and linked with each other. Apart from these two spatial planning teams only two other teams (33 and 50) are directly related to these academic teams. The consulting firm Infocharta (team 33) was founded by the head of

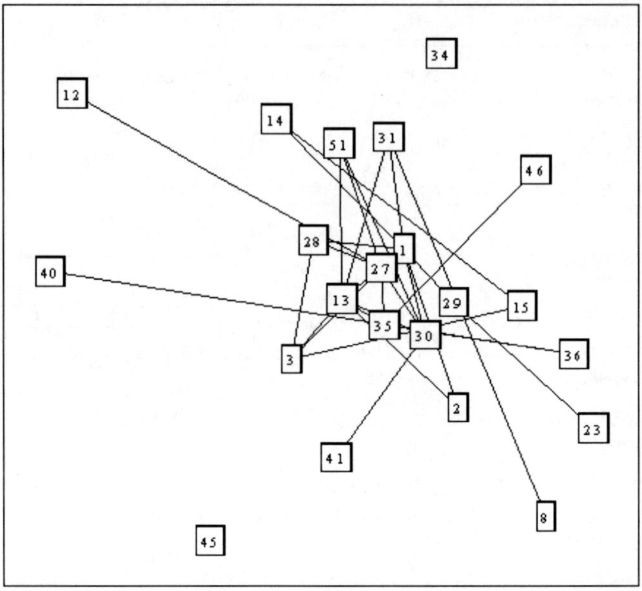

Graph 5.10 Teams which share a surveying engineering disciplinary background based on the Euclidian distances between them.

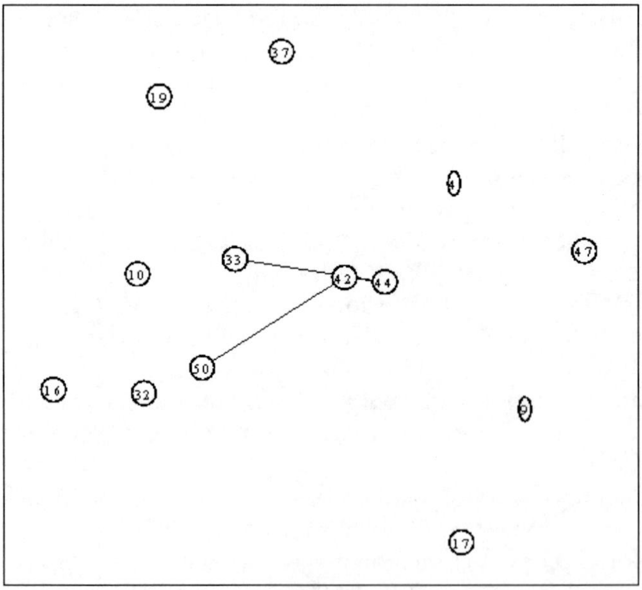

Graph 5.11 Teams which share a spatial planning disciplinary background based on the Euclidian distances between them.

team 44, while the head of the Technical Services (team 50) at the municipality of Rodhos has attended many of the GIS seminars organised by team 42.

Last, Graph 5.12 shows the third group of teams that share one of the other disciplinary backgrounds related to GIS in Greece. As might be expected, the Greek ESRI vendor (Marathon Data Systems, team 21) is the reference point for this group, providing connectedness to most of its members. The only isolates within this group are other GIS software vendors, namely Intergraph Hellas (team 22), Omas-MapInfo (team 25) and Star (team 26). Like the spatial planning group, only a small minority of the 'other' teams (three out of 16: Marathon Data Systems, Intergraph Hellas and the academic team based at the Department of Environmental Studies at the University of the Aegean) belong to the inner circle of the Greek GIS community. The main reason for the lack of direct linkages among the teams of this heterogeneous social group is that they follow various traditions of practice related to GIS and their members maintain multiple memberships with technological communities ranging from computer science to human geography and agriculture and forestry.

5.9 Summary

This chapter explored the origins and formation of the critical mass of the Greek GIS community. Section 5.2 identified and discussed the size and composition of the functional components/institutional groups of the Greek

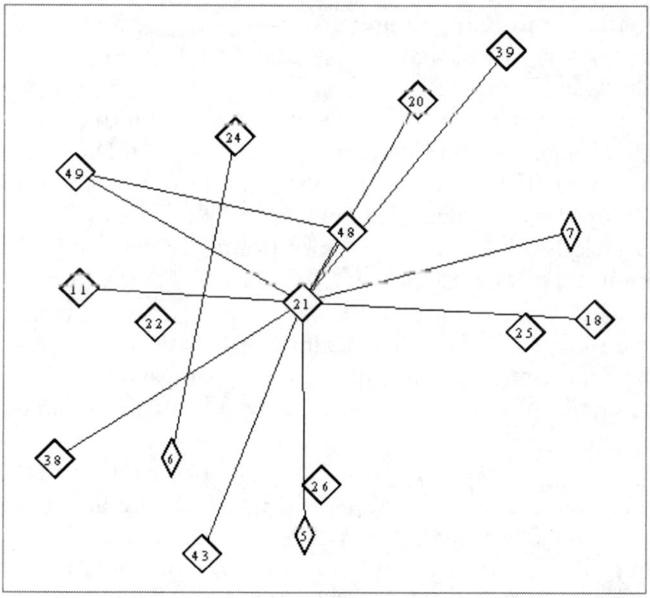

Graph 5.12 Teams which share the 'other' disciplinary backgrounds based on the Euclidian distances between them.

GIS community based on 51 GIS teams in central government agencies, municipalities, utility organisations, university laboratories and private sector firms (see Table 5.1). Academics form the largest group, representing 35 per cent of the teams in the Greek GIS community, while GIS vendors and consultants are the next largest group with almost 30 per cent. Various issues were also pointed out regarding the overlapping and shifting membership of GIS teams over time. Section 5.3 highlighted the interdependencies in the 'triple helix' of institutional groups of GIS teams forming the Greek GIS community, as well as a number of essentially complementary areas of interest between the main components of GIS technology and the main components/institutional groups of the Greek GIS community.

Section 5.4 showed that there were four stages – cosmopolites, innovators, early adopters and early majority – in the development of the critical mass of the Greek GIS community from the mid-1980s to the mid-1990s. The first GIS teams in government, universities and the private sector moved in parallel in the late 1980s and through the early 1990s with respect to the adoption and implementation of GIS technology. Before these innovators a handful of cosmopolitan GIS experts studied and socialised abroad in North America and Europe from the late 1960s to the early in 1980s. GIS technology is a rather recent phenomenon in Greece, since the first GIS software licence was installed in 1986. Most organisations started adopting GIS software around 1990. Since 1990 the adoption of GIS has accelerated significantly, with organisations throughout the country mainly investing in hardware and GIS software. By 1994–5 pilot GIS projects had been developed in all groups of the triple helix of university–industry–government institutions and the necessary critical mass for sustaining GIS diffusion in Greece had been reached (see Figures 5.1 and 5.2).

Sections 5.5 to 5.8 explored with SNA and visualisation techniques the findings of the earlier analysis based on Figure 5.3 of GIS teams and heterogeneous linkages connecting these teams throughout Greece. Specialised computer software was used to visualise and analyse the patterns of these socio-economic and knowledge linkages and the positioning of the teams of the Greek GIS community. Using Euclidian distances as a measure of structural equivalence (see Appendix) the patterns of direct linkages were explored with reference to two social constructs: institutional and disciplinary arrangements. The inner, middle and outer circles of the Greek GIS community were also defined based on structural equivalence and the time GIS software was adopted.

Section 5.6 explored the prominence of specific teams such as the ESRI vendor, and relevant social groups that shaped GIS technology and the Greek GIS community in its early critical stages of development. The analysis of cohesive subgroups and centrality highlights the prominence of team 21 (Marathon Data Systems) as well as the central role of a small minority of surveying engineering teams. The 31 cliques identified indicate the extent to which surveying engineering teams outnumber any other cohesive subgroup within the Greek GIS community. The findings of the various centrality measures suggest that a distinction should be made between 'monomorphic' opinion leaders such as team 21 and 'polymorphic' opinion leaders such as team 13 (Laboratory of Cartography at the NTUA). The latter has not only made a significant contribution to the development of the Greek GIS community in the area GIS software but it has also critically influenced the setting up of new institutions for geographic information handling and the teaching of GIS-related courses. This supports the argument put forward by Rogers and Kincaid (1981) that opinion leadership with respect to technological innovations such as GIS is more a systemic and structural characteristic than an individual one. Hence a significant finding of this analysis is that it shifts the debate about the roles in GIS innovation from individual 'trees' to the 'forest' of GIS linkages.

With respect to the two social constructs, the Euclidian distances between the 51 teams indicate that there is a great deal of heterogeneity in the inner circle of the Greek GIS community in terms of the institutional setting (see Section 5.7), as a handful of teams from the 'triple helix' seem to transcend institutional boundaries, having developed similar patterns of connections in collaboration and competition for GIS adoption and implementation. In contrast to the findings for the institutional setting, the findings from the analysis of the disciplinary backgrounds (see Section 5.8) highlight the dominance of teams with a surveying engineering background. Two-thirds (68 per cent) of the inner circle of the Greek GIS community share a surveying engineering background, while the rest are divided between the teams with a spatial planning and one of the other disciplinary backgrounds related to GIS in Greece (see Table 5.6). As a result of the predominance of the surveying engineering tradition of practice in the emerging structure of the Greek GIS community since the early 1990s, applications have systematically focused on cadastral and topographic databases channelling most of the EU resources in the last decade or so to this technological community and tradition of practice (Assimakopoulos 1997c).

6 A regional semiconductor community and academic entrepreneurship in Silicon Valley

6.1 Introduction

This chapter further narrows down the focus of inquiry to a regional rather than national scale. On one hand, it focuses on the origins and emergence of the technological community of semiconductor firms in Silicon Valley (SV), California, and in particular its unique horizontal and entrepreneurial structure. The latter is worth studying in its own right, but also because it has served as a model for subsequent technological communities in SV. Furthermore this chapter explores university–industry links through a network analysis of leadership in professorial entrepreneurship at the departments of electrical engineering and computer science at Stanford University and the University of California at Berkeley (UCB). It therefore highlights the significance of academia in the continuing success of SV through successive generations of ICT RTD work, from semiconductors in the 1960s and 1970s to personal computers in the 1980s, and Internet-related mobile technologies in the 1990s and early 2000s.

More specifically, Section 6.2 describes the origins of the first semiconductor firms in SV and introduces a genealogy chart of SV semiconductor firms published by the Semiconductor Equipment and Materials International (SEMI). It also tells the story of how William Shockley's transistor laboratory 'gave birth' in 1957 to Fairchild Semiconductor when eight of his top employees left out of frustration with his management style and created Fairchild Semiconductor, and how these 'serial' entrepreneurs subsequently set up 'Fairchildren' spin-off firms such as Intel in 1968. Section 6.3 takes stock of network analysis and visualisation techniques, and using the SEMI genealogy chart it analyses the inter-personal networks of co-founders of the semiconductor firms in SV, as well as the networks of firms that these founders came from. Section 6.4 sheds additional light on the evolution of the community of semiconductor firms and the centrality of Fairchild in spreading an entrepreneurial and horizontal community-like culture in SV through successive generations of 'Fairchildren' spin-offs, up to the late-1980s. As was anticipated, the analysis of centrality and prominence of founders and semiconductor firms fuelling the explosive growth of semiconductor RTD work in SV brings to light the usual suspects: the 'traitorous eight' founders of Fairchild as well as Intel and Hewlett-Packard (HP), together with a relatively unknown firm outside the semiconductor community, Intersil Corporation.

Section 6.5 shifts the focus on the entrepreneurial networks of electrical engineering (EE) and computer science (CS) professors at Stanford and UCB, analysing not only their corporate involvement with regard to co-founder relations, but also shedding light on their director and advice networks with firms in SV and beyond. Overall, compared with their UCB colleagues, Stanford professors demonstrate a higher level of leadership and corporate involvement with respect to the founder and director relationships, though UCB professors have more pronounced advice networks among themselves and with firms in the region and beyond. Last, Section 6.6 is the chapter summary.

6.2 The SEMI genealogy chart and origins of the community

Many SV firms have a semiconductor 'genealogy chart' hanging in their foyers, tracing their roots back to Fairchild and Shockley in the late 1950s. Don Hoefler (1971) first put together this SV 'genealogy chart' which was subsequently updated by the trade association SEMI (www.semi.org). This chart identifies all 372 founders and 129 firms started up in SV from 1957 to 1986. It is also illustrative of the entrepreneurial and collaborative spirit that has generated so many new start-ups in the area and made SV the centre of the worldwide high-tech industry for the past few decades. Instead of staying in a large successful firm, engineers and other employees of SV firms have traditionally preferred to launch their own start-ups, with the hope of striking it rich with the equity return. Thus, analysis of this chart provides insights into the entrepreneurial culture and 'horizontal' network structure of the semiconductor community in SV. A multi-level network analysis of this technological community is presented below. But first it is worth telling the story of how the first semiconductor firms came to be in SV and the origins of the semiconductor revolution and technological tradition of practice that was set off by a handful of men in the late 1950s.

William J. Shockley, John Bardeen and Walter Brattain of Bell Laboratories in Murray Hill, New Jersey, demonstrated the first successful transistor on 23 December 1947. These three inventors eventually won a Nobel Prize for this achievement, but only Shockley was determined to capitalise on it. He left Bell Laboratories in 1954 for New England to become a consultant for Raytheon Corporation with the hope of establishing a semiconductor firm (Riordan and Hoddeson 1997). However, after Raytheon refused to guarantee him US$1 million in seed money over three years, he left after only one month, and headed back to his hometown, Palo Alto in California, where his mother still lived. There, with the backing of Arnold Beckman, the president of Beckman Instruments, and the encouragement of Frederick

Terman, the engineering dean at Stanford University, he started his own company, Shockley Transistors, in 1955.

By most accounts Shockley was a genius for spotting and recruiting talent, but he was not as adept at managing it. Within two years of its founding, Shockley Transistors was experiencing severe internal turmoil, largely due to Shockley's bizarre management style. According to Wolfe (1983), Shockley published his employees' salaries on a bulletin board, started having them rate one another, and even instigated lie detector tests when he suspected that someone was sabotaging his project. Above all Shockley's decision to concentrate on four-layer diodes rather than the product implied by the company's name led a group of his employees to approach Beckman with a plan that would grant Shockley emeritus status in the company but remove him from the day-to-day operations of the firm. However, Shockley vetoed the plan. As a result eight of his employees left in 1957 in order to found Fairchild Semiconductor. Shockley branded these defectors the 'traitorous eight'.

> These men, Sheldon Roberts, Eugene Kleiner, Jean Hoerni, Gordon Moore, Jay Last, Victor Grinich, Julius Blank, and Robert Noyce, were highly creative and independent-minded scientists with impeccable credentials. Five had PhDs in the physical sciences and had done noted research work in spectroscopy, metallurgy, and solid state physics. Noyce, a solid state physicist from MIT, was the only one with a strong semiconductor background. Like Noyce and Last, Roberts, a metallurgist, was an MIT PhD and Moore, a physical chemist from Cal Tech, had worked on the spectroscopy of hot gases for the missile re-entry program at Johns Hopkins. Finally, Hoerni was a physicist from Switzerland, with two PhDs from Oxford, and the University of Geneva. The group also had three engineers. Kleiner, an Austrian émigré, had studied mechanical and industrial engineering in the US before designing cigar-making machinery at the American Shoe Foundry Company and making relays at Western Electric. Blank, a mechanical engineer, had held a variety of technical and engineering positions at Babcock & Wilcox and Western Electric. Finally, Grinich, the group's only electrical engineer, had received a doctorate in circuit theory from Stanford University. At the Stanford Research Institute, he designed transistor circuits for color television and the ERMA computer, the first computing machine for banking applications.
>
> (Lecuyer 2000: 160)

Shockley Transistors never recovered from the departure of these eight men. It did, however, hang on until 1968 when Beckman sold it to Clevite, which in turn sold it to ITT, which eventually shut it down. Fairchild, on the other hand, met with almost immediate success. It initially manufactured high-frequency transistors based on techniques developed at Shockley. It was turning a profit by the end of 1958. The timing of its founding could not have been better.

> By 1957 there was sufficient demand from manufacturers who merely wanted transistors instead of vacuum tubes, for use in radios and other machines, to justify the new operation. But it was also in 1957 that the Soviet Union launched Sputnik I. In the electronics community the ensuing space race had the effect of coupling two new inventions – the transistor and the computer – and magnifying the importance of both.
>
> (Wolfe 1983: 358)

Robert Noyce, one of the 'traitorous eight', helped develop the first integrated circuit (IC), which eventually turned Fairchild into one of SV's most profitable firms. Strictly speaking, Noyce was not the first to develop an IC. This honour belonged to Jack Kilby of Texas Instruments who was awarded the Nobel Prize for his accomplishment in 2000 (Berlin 2005). But six months after the announcement of Kilby's invention, Noyce with his R&D team created a similar circuit except that it was made of silicon and used a unique insulating technique, the planar process developed by Jean Hoerni, another of the 'traitorous eight'. Noyce's device turned out to be more cost-efficient and more practical to make than Kilby's. It eventually became the community standard, triggering a technological revolution in ICs. It also helped turn Fairchild into the new semiconductor community's leader by capitalising on the 'presumptive anomalies' (Constant, 1987) arising from the invention of the planar process for manufacturing ICs and put SV on a path that would eventually lead to its dominance in the high-technology community worldwide. According to Berlin (2005: 109–11):

> Whereas Kilby at Texas Instruments asked 'how can I build an integrated circuit?' Noyce wondered 'how can this planar process be used?' (Noyce: 'I was trying to solve a production problem. I wasn't trying to make an integrated circuit.') Noyce was thus focused on production from the beginning; with this intellectual launch pad, it would have been difficult for him to consider seriously any device that would not have been mass produced … Noyce filed his integrated circuit patent on July 30, 1959 … The man who bore the brunt of moving the integrated circuit to the third stage, from 'it's possible' to 'it's finished,' was Jay Last.

In spite of the impact that ICs have had in SV's development as a technological leader, Fairchild's importance for SV transcends this particular technological breakthrough as its founders, the 'traitorous eight', triggered a new technological tradition of practice (Constant 1987) and technological community of semiconductor firms. Under Noyce's leadership Fairchild also contributed a vision of management that explicitly rejected the hierarchical East Coast corporate culture (Saxenian 1996). This was perhaps in part in reaction to Shockley's authoritarian management style, but regardless of its roots, this new approach spread as employees from Fairchild left to start their own firms. Researchers often highlight two components of the unique organisational culture found in SV firms: a community-like atmosphere within the firms and an entrepreneurial spirit that spawns many start-ups. First, the management style of SV firms is in stark contrast to that of East Coast firms. According to Wolfe (1983: 360):

> Corporations in the East adopted a feudal approach to organization, without even being aware of it. There were kings and lords, and there were vassals, soldiers, yeomen and serfs, with layers of protocol and perquisites, such as the car and driver, to symbolize superiority and establish the boundary lines ... Fairchild Semiconductor needed a strict operating structure, particularly in this period of rapid growth, but it did not need a social structure ... Noyce rejected the idea of a social hierarchy at Fairchild ... Everywhere the Fairchild émigrés went, they took the Noyce approach with them. It wasn't enough to start up a company; you had to start a community, a community in which there were no social distinctions, and it was first come, first served in the parking lot, and everyone was supposed to internalize the common goals. The atmosphere of the new firms was so democratic, it startled businessman from the East.

Fairchild also contributed to the development of a second key organisational culture of the Valley that encourages entrepreneurial risk-taking: a culture that Saxenian argues played an important role in SV's success throughout the 1980s and 1990s. According to Saxenian (1996), in places where failure is frowned upon, it is unlikely that very many people will engage in high-risk activities such as business start-ups. As an example she points to the relatively risk-averse culture of the high-technology industrial region of Route 128 in Massachusetts that gave birth to few successful start-ups during the same period. SV's culture, on the other hand, supports risk-taking and this has led to a large number of start-ups, and such entrepreneurial activity can be traced back to the traitorous eight who left Shockley Transistors to form Fairchild Semiconductor.

If such entrepreneurial risk-taking had stopped there, then perhaps the start-up culture would not have taken roots in SV, but it did not. When Fairchild was only 18 months old, its first general manager Ed Baldwin left to set up Rheem Semiconductor and also took with him 10 key Fairchild employees. Fairchild sued Baldwin and Rheem, claiming that he had stolen Fairchild's process manual. The court, however, was not terribly impressed with Fairchild's argument since Fairchild essentially did to Shockley what they claimed Rheem did to them. In the end the two firms settled the case out of court whereby Rheem agreed to pay Fairchild US$70,000 and to refrain from using one of Fairchild's proprietary processes. However, while the settlement did not cripple Rheem financially, it did in other ways, and after only 30 months in business it was acquired by Raytheon, the company which six years before had refrained from funding Shockley's business plan.

When relations between Fairchild and its parent firm, Fairchild Camera and Instrument Corporation, became strained, four of its original founders – Hoerni, Kleiner, Last and Roberts – pulled out in 1961 to form Amelco (now Teledyne Semiconductor). In the same year several key Fairchild employees left to found Signetics. Undoubtedly, however, the most important spin-off occurred in 1968 when Robert Noyce, along with Gordon Moore and Andy Grove, left to start Intel. By 1986 31 semiconductor firms could directly trace their ancestry to Fairchild. And these do not include those that can indirectly trace their ancestry to Fairchild whose founders came from firms such as Intel that were founded by Fairchild employees. In the next sub-section we therefore shed light in the centrality of Fairchild founders for the semiconductor community in SV in the early critical stages of its development by looking into both the founders' networks and the networks of firms that these founders came from.

6.3 Founders and founders' previous firms

The SEMI 'genealogy chart' identifies all 372 founders of SV semiconductor firms and the 105 firms they came from during the late 1950s to the late 1980s. Two firms are connected in the SEMI chart when one of them is founded by someone from the other. While the chart portrays the community's evolution over time, the connections between firms, let alone individuals, are very difficult to see and study because of the volume and complexity of linkages. More importantly, it is difficult to comprehend the processes of spin-off creation and the pattern of collaborative efforts among individual founders fuelling the creation of critical mass (Rogers 2003) and the development of the semiconductor community in SV. The use of social network analysis and visualisation software, such as Ucinet (Borgatti, Everett and Freeman 2002), Pajek (Batagelj and Mrvar 2004; De Nooy *et al.* 2005) and Mage (Richardson and Presley 2001) provide a powerful methodology which fulfils both objectives. The network analysis and visualisation of inter-personal and inter-firm networks is not only essential in understanding the emergence of a collaborative and entrepreneurial culture in SV, first initiated with the start-up of Fairchild, and subsequently spread throughout the region with dozens of 'Fairchildren' firms, it also enables an increased understanding in terms of identifying the community's central founders across the generations of spin-offs. Note that a tie exists between two co-founders when these individuals have co-founded a firm.

Graph 6.1 visualises the network of SEMI co-founders using Ucinet's (Borgatti, Everett and Freeman 2002) Netdraw spring embedding layout based on geodesic distances as a measure of proximity between nodes, i.e. founders. Spring embedding algorithms are based on an assumed attraction

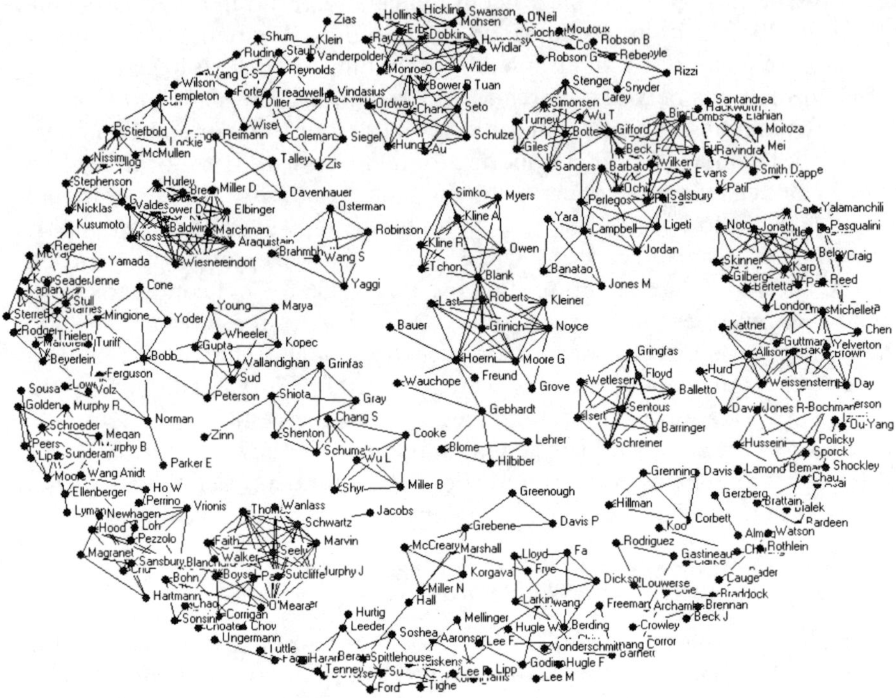

Graph 6.1 SEMI founders, 1957–86.

between adjacent nodes, allocating nodes to a two-dimensional space according to a measure of proximity such as the shortest path, or geodesic distance between any two nodes in the underlying socio-matrix. As can be seen from Graph 6.1 there are a number of dense components with co-founders clustering in distinctive subgroups. Overall, however, the visualisation in Graph 6.1 is rather intricate as there are too many nodes, subgroups and ties for identifying any clear pattern in the co-founding activity of these individuals.

It is worth noting that William Shockley is positioned at the edge of the community in the right-hand bottom half of Graph 6.1, as he did not co-found or collaborate further in the development of this technological community. Table 6.1 summarises the degree and betweenness centrality (see Appendix) scores for the top 20 founders in Graph 6.1. As expected, about two-thirds of the top 20 centrality scores both in terms of number of ties (degree) or/and power (betweenness) are Fairchild founders and employees who have co-founded dozens of semiconductor firms.

In a more detailed view Graph 6.2 shows the large subgroups/components with more than nine co-founder members, who include the co-founders with the highest centrality scores, plus the name and year of establishment of the start-up firms. It is worth highlighting the cases of the most central SEMI co-founders: D Miller, Hoerni, Gifford and Blank (see Table 6.1). D Miller,

Table 6.1 Top 20 centrality scores of SEMI founders

Node	*Founder*	*Degree centrality*	*Node*	*Founder*	*Betweenness centrality*
218	Miller D	6.739	147	Hoerni	0.159
118	Gifford	4.043	35	Blank	0.131
147	Hoerni	3.774	118	Gifford	0.082
35	Blank	3.504	116	Gebhardt	0.079
155	Hurley	3.235	37	Bobb	0.068
354	Wiesner	3.235	218	Miller D	0.048
7	Araquistain	3.235	220	Mingione	0.029
14	Baldwin	3.235	34	Blanchard	0.029
50	Breene	3.235	54	Campbell	0.026
333	Valdes	3.235	235	Norman	0.016
182	Koss	3.235	237	Noyce	0.014
349	Weindorf	3.235	225	Moore G	0.014
43	Bower D	3.235	285	Schwartz	0.013
95	Elbinger	3.235	252	Perlegos	0.012
205	Marchman	3.235	13	Baker	0.009
266	Roberts	2.695	2	Allison	0.009
187	Last	2.695	124	Grebene	0.009
285	Schwartz	2.695	291	Shiota	0.006
174	Kleiner	2.695	135	Hall	0.003
318	Sutcliffe	2.426	206	Marshall	0.003
237	Noyce	2.426			

an ex-Fairchild employee, scores the highest degree centrality having co-founded both Rheem Semiconductor in 1959 and Sensym in 1982 (see right-hand upper half of Graph 6.2). Ed Baldwin is also notable among the Rheem people, though his betweenness centrality is zero as he does not stand on the shortest path of any other two co-founders. Note also that Sensym was founded by employees of National Semiconductor, one of the 'Fairchildren' firms, which was also founded in 1967 by ex-Fairchild employees. Hoerni, the inventor of the planar process, occupies the most powerful position (highest betweenness centrality) in the main component/subgroup of the co-founder network.

Graphs 6.3 and 6.4 show the main component of the SEMI co-founders' networks with the 'traitorous eight' Fairchild co-founders at its centre. The difference between Graph 6.3 and 6.4 is that the latter is a two-mode network showing two sets of actors, i.e. the network of founders and the firms founded. Hoerni, Last, Roberts and Kleiner are connected with thicker ties in Graph 6.3 as they also co-founded Amelco in 1961 (see also Graph 6.4). Blank also scores very highly (second betweenness centrality and fourth degree centrality (see Table 6.1) as he is one of the 'traitorous eight' and also co-founded Xicor with employees of 'Fairchildren' firms Intersil and Intel in 1978. Last but not least, another ex-Fairchild employee, Gifford, scores second in degree and third in betweenness centrality as he co-founded both Advanced Micro Devices (AMD) in 1969 and Maxim Integrated Products in

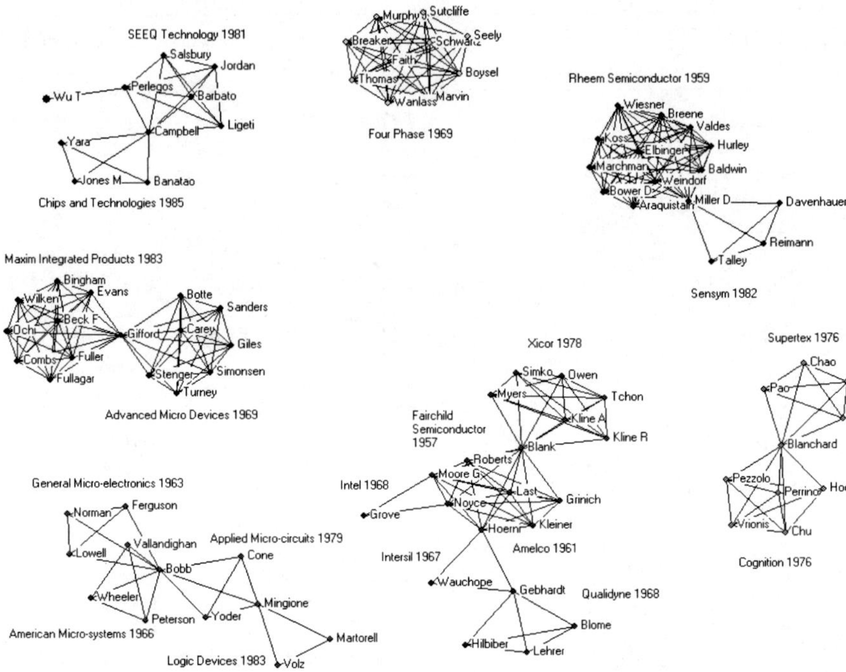

Graph 6.2 SEMI founders' main components ($N > 9$).

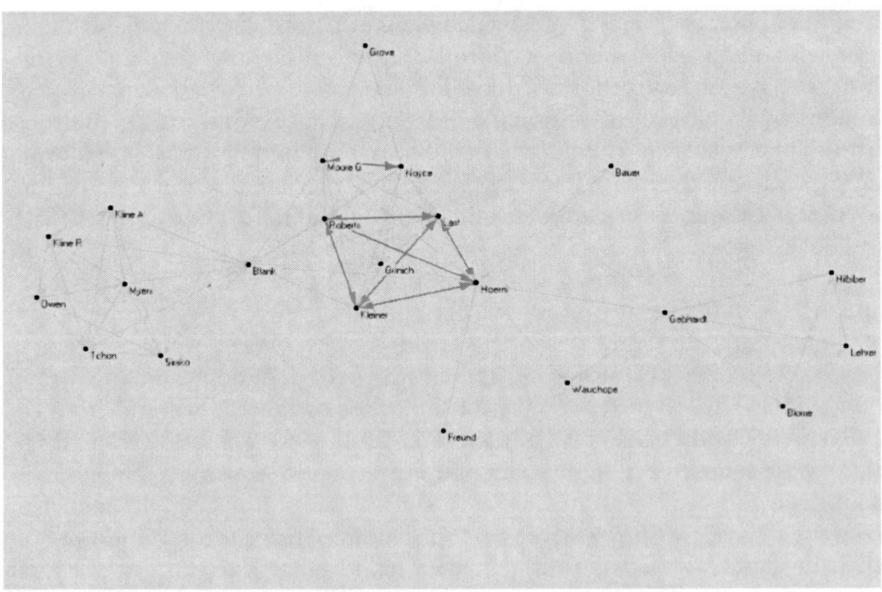

Graph 6.3 SEMI founders' main component ($N = 22$) – 1 mode.

1983. Again AMD is one of the 'Fairchildren' firms, with all its founders coming from Fairchild, and Maxim Integrated Products was co-founded with others not only from Fairchild, but also from 'Fairchildren' firms Intersil and Amelco/Teledyne. Gordon Moore has elaborated on this phenomenon by pointing out that:

> In any case, integrated circuits, MOS transistors, and the like proved too rich a vein for a company the size of Fairchild to mine, resulting in what came to be termed the 'Silicon Valley effect'. At least one new company coalesced around and tried to exploit each new invention or discovery that came out of the lab. Notwithstanding spin-offs in all directions, there was still plenty to keep Fairchild Semiconductor growing as rapidly as it could.
>
> (Moore 1996: 167)

These findings beg for further analysis of all founders' previous affiliations with firms in SV and beyond. The SEMI 'genealogy chart' shows 105 affiliations of the individual founders to a previous firm. Graph 6.5 visualises the network of SEMI founders' previous firms with Pajek. A tie exists between two firms in this graph when at least two individuals who come from these firms have co-founded a third firm on the chart. There is a large

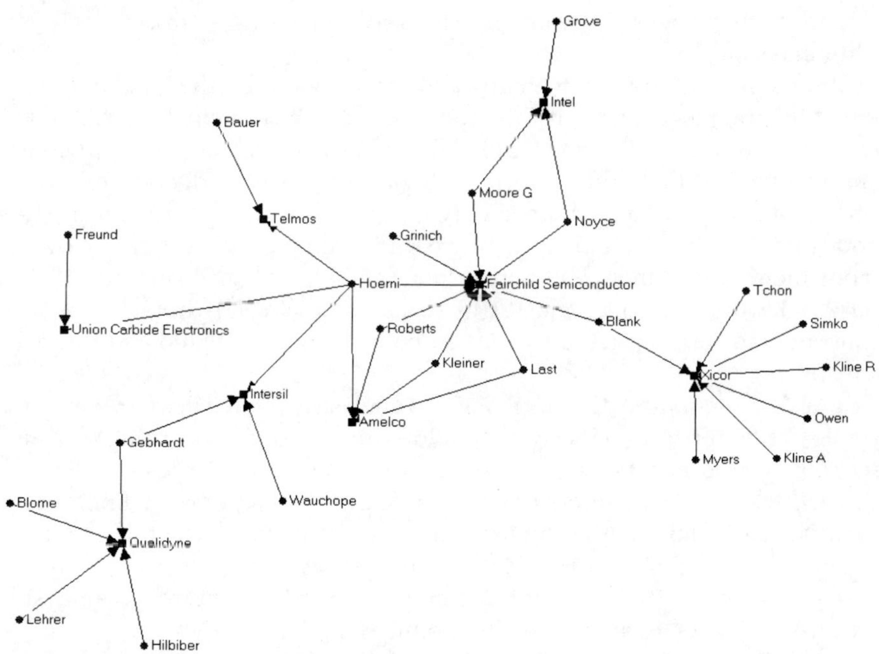

Graph 6.4 SEMI founders' main component ($N = 22$, $M = 8$) – 2 mode.

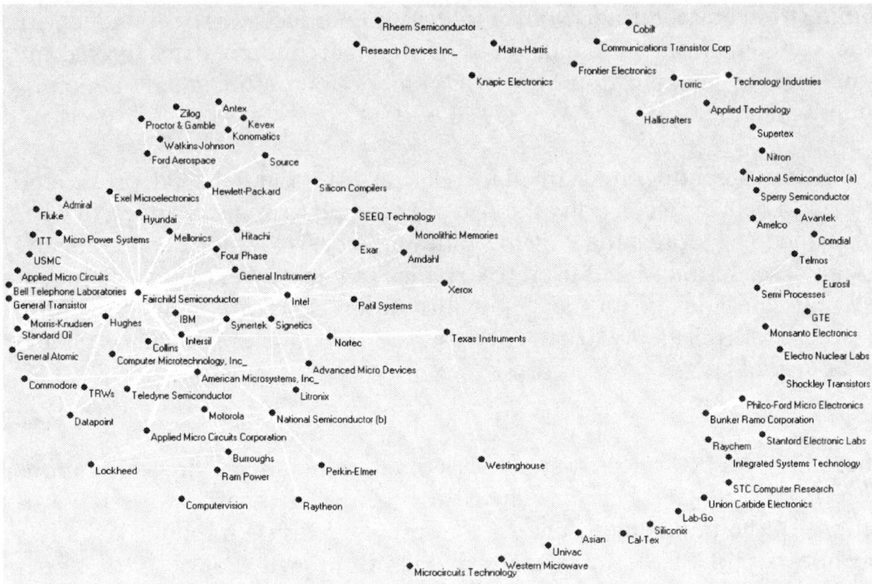

Graph 6.5 SEMI founders' previous firms, 1957–86.

main component/subgroup with Fairchild and Intel at its centre in Graph 6.5, and many isolates or dyads of firms positioned in a semi-circle on its right-hand side.

Subsequently in Graph 6.6 Netdraw is used to visualise the main component of the founders' previous firm network using the principal components layout. As can be seen from Graph 6.6 the network of founders' previous firms can be divided into two parts. Fairchild single-handedly occupies the right-hand side, providing connectivity to the whole graph, and in top left-hand quarter a large number of firms with prominent nodes are shown, among them Intel, American Microsystems and Intersil, while in the bottom quarter there is a smaller cluster of firms where those who co-founded semiconductor firms together with their colleagues in Fairchild and Intersil came from.

Table 6.2 summarises the results of the analysis of centrality measures in Graphs 6.5 and 6.6. Obviously Fairchild is the most central firm, both in terms of degree and power in the network of founders' previous firms, followed by Intel, HP, Intersil and American Microsystems. A limitation that arises from this analysis, however, is related to time, as the time dimension in the original chart has been lost, and the evolution of the founders' networks seems difficult to trace using the available network analysis software. In order to illustrate this point, Graph 6.7 shows the network of SEMI-founded firms by the founders' previous firm affiliation. This two-mode (firm-by-firm) graph shows clearly the dominance of the firms

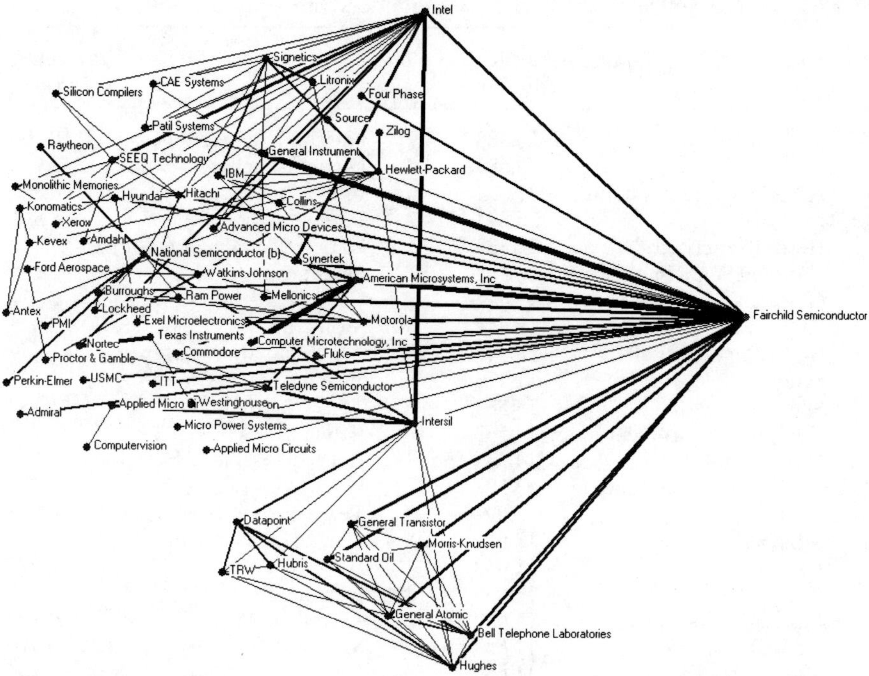

Graph 6.6 SEMI founders' previous firms – principal component analysis.

identified above and in Table 6.2 but as there is no chronology built into the data analysis, firms appear twice: for example, Intel as the firm where the founders of LSI Logic came from, and Fairchild as the firm where the founders of Intel came from.

Assimakopoulos, Everton and Tsutsui (2003) have dealt with this limitation in their recent study of the evolution of the semiconductor community in SV, highlighting six points in time covered by the genealogy chart: 1960, 1965, 1970, 1975, 1980 and 1986, tracing the cumulative change in the community's network structure as well as its central firms across these periods (see also Everton 2004). Six socio-matrices (one for each point in time) were created in which a tie exists between two firms if individuals from the two firms joined forces to found a third.

The underlying argument is that over time, as the number of co-founding ties between firms grew and resulted in an increasingly denser network among co-founders, Fairchild's collaborative and entrepreneurial culture spread throughout SV, as the necessary critical mass of co-founder relationships was built up for diffusing the community-like organisational culture first initiated in Fairchild and subsequently spread throughout the Valley through dozens of 'Fairchildren' spin-offs. The increasing number of start-up semiconductor firms in itself is also evidence that the entrepreneurial spirit became increasingly prevalent and took root in SV. In this sense, the

Table 6.2 Top 20 centrality scores of SEMI founders' previous firms

Node	*Firm*	*Degree centrality*	*Node*	*Firm*	*Betweenness centrality*
29	Fairchild Semiconductor	120.388	29	Fairchild Semiconductor	18.790
46	Intel	43.689	39	Hewlett-Packard	8.977
5	American Microsystems	37.864	46	Intel	8.887
47	Intersil	36.893	47	Intersil	8.591
39	Hewlett-Packard	22.330	65	National Semiconductor (b)	4.840
65	National Semiconductor (b)	20.388	81	Signetics	2.904
35	General Instrument	19.417	67	Nortec	2.322
42	Hughes	18.447	63	Motorola	2.045
12	Bell Telephone Labs	18.447	5	American Microsystems	1.450
90	Synertek	17.476	8	Applied Micro Circuits Corp.	1.180
81	Signetics	15.534	94	Texas Instruments	1.180
22	Computer Microtechnology	14.563	40	Hitachi	0.747
24	Datapoint	14.563	12	Bell Telephone Labs	0.661
92	Teledyne Semiconductor	13.592	42	Hughes	0.661
67	Nortec	13.592	92	Teledyne Semiconductor	0.615
78	SEEQ Technology	12.621	35	General Instrument	0.594
44	IBM	11.650	44	IBM	0.432
34	General Atomic	11.650	14	Burroughs	0.340
62	Morris-Knudsen	11.650	2	Advanced Micro Devices	0.199
36	General Transistor	11.650	78	SEEQ Technology	0.114
86	Standard Oil	11.650	53	Litronix	0.104
40	Hitachi	9.709	90	Synertek	0.081
8	Applied Micro Circuits Corp.	9.709	79	Semi Processes	0.019
63	Motorola	8.738			
94	Texas Instruments	8.738			

next sub-section sheds additional light on the network 'horizontal' structure of the co-founded firms, small and large alike, that still gives SV part of its regional advantage (Saxenian 1996).

6.4 Firms founded and the centrality of Fairchild Semiconductor

The networks of co-founded semiconductor firms are presented next using the Mage network visualisation software (Richardson and Presley 2001), in conjunction with the ranking of firms according to their centrality measures calculated with Ucinet (Borgatti, Everett and Freeman 2002) clarifying who the central and powerful actors in the community were in each period from the late 1950s to the late 1980s. It is worth noting from the outset that these graphs are independent of each other and the location of each node, i.e. firm, changes with each graph.

In the first period, up to 1960 (see Graph 6.8), the development of the semiconductor community is at a very early stage and no clear central actor has emerged. The community network consists of nine actors and is rather

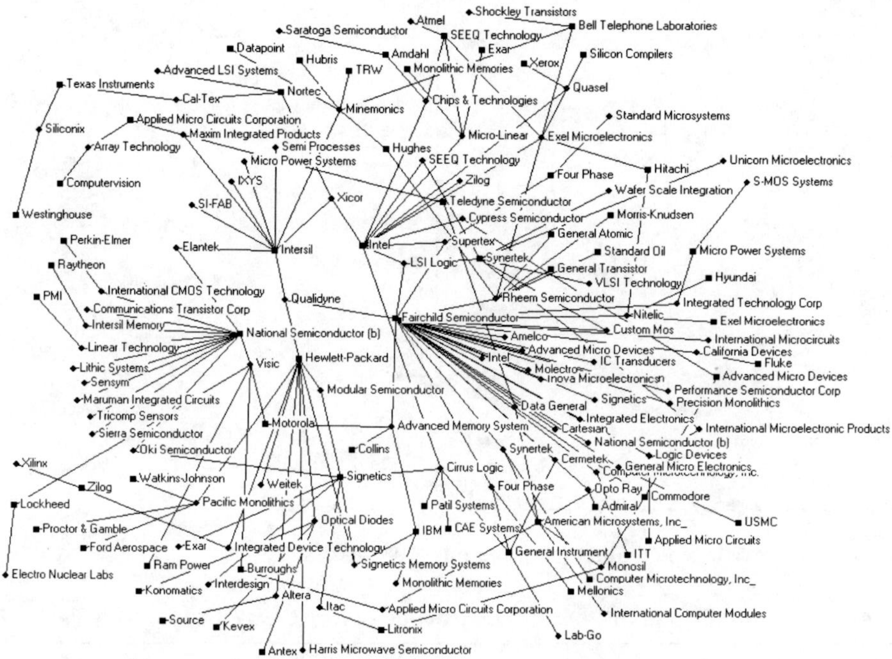

Graph 6.7 SEMI founders' previous firms – main component.

symmetric, with all but two of the nine firms being connected with others. The two that are not are Shockley Transistors and Sperry Semiconductor. Their isolation in Graph 6.8 indicates that employees from these two firms did not co-found new firms with employees from other firms in the graph (see also Graph 6.1). Hence, the image shows no tie between them and the other seven firms in the graph. It is fascinating, however, that even at this very early stage in the community, employees from seven firms joined together in the founding of a new firm (Rheem). Table 6.3 shows that, of the network's nine actors, seven share the same measures of degree centrality. This reflects the fact that all seven are connected similarly with one another and no single actor stands out as the central actor. This fact is perhaps better captured by the betweenness centrality measures, all of which equal zero, indicating that no single actor at this point in time is more 'powerful' or influential than the others.

By 1965 the network had grown to 14 firms. As Graph 6.9 illustrates, in 1965 SV's semiconductor community was still in an embryonic stage and appeared to have changed very little since 1960. The seven firms that were connected with each other in 1960 remain the community's central sector, but an eighth company has joined the group: namely, the United States Marine Corps (USMC). As Graph 6.9 indicates, however, the USMC is not connected with the other seven but only to one (Fairchild). What in fact

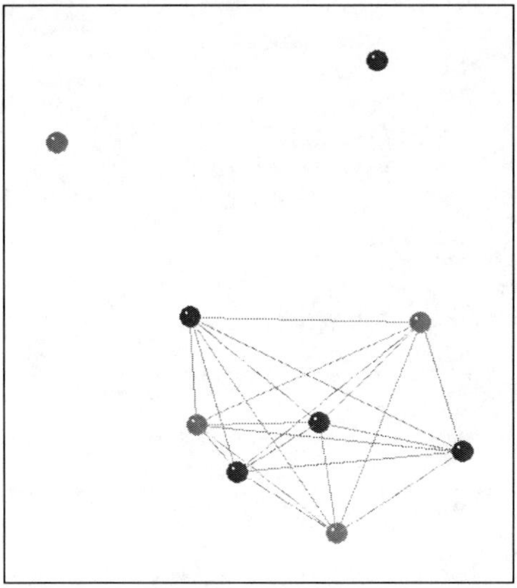

Graph 6.8 SEMI semiconductor community, 1947–60.

Table 6.3 Centrality ranking of SV's semiconductor firms, 1947–60 ($N = 9$)*

	Degree centrality	*Betweenness centrality*
1. Bell Telephone Laboratory	75.00	0.00
1. Fairchild Semiconductor	75.00	0.00
1. General Atomic	75.00	0.00
1. General Transistor	75.00	0.00
1. Hughes	75.00	0.00
1. Morris-Knudsen	75.00	0.00
1. Standard Oil	75.00	0.00
8. Shockley Transistor	0.00	0.00
8. Sperry Semiconductor	0.00	0.00

Note
* The ranking is based on the degree centrality score for all periods.

occurred is that in 1963 three employees from Fairchild and one from the USMC joined together to found General Micro Systems. Because of this co-founding, by 1965 Fairchild began to show signs of becoming the semiconductor community's central actor, at least in terms of the founding of new semiconductor firms. Graph 6.9 also shows the presence of a tie between two firms apart from the community's central sector. This tie reflects the co-founding of Siliconix by employees from Texas Instruments and Westinghouse.

Table 6.4 underscores this point further. In 1965 Fairchild had a slightly higher normalised degree centrality (53.85) than the rest of the other actors in the network. It was also the only actor to register a betweenness centrality

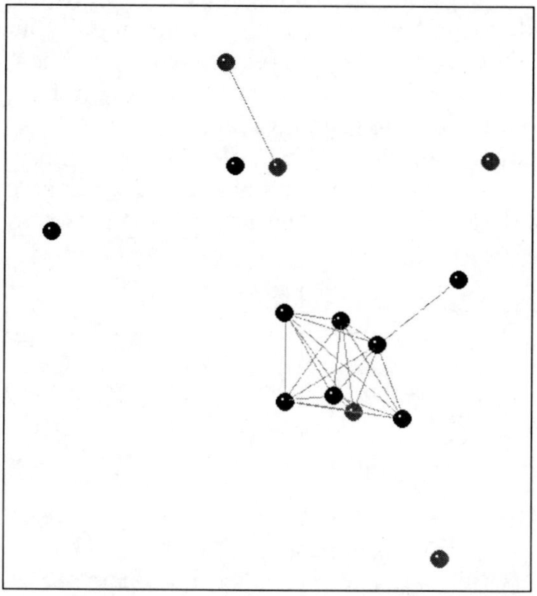

Graph 6.9 SEMI semiconductor community, 1947–65.

Table 6.4 Centrality ranking of SV's semiconductor firms, 1947–65 ($N = 14$)

	Degree centrality	*Betweenness centrality*
1. Fairchild Semiconductor	53.85	7.69
2. Bell Telephone Laboratory	46.15	0.00
2. General Atomic	46.15	0.00
2. General Transistor	46.15	0.00
2. Hughes	46.15	0.00
2. Morris-Knudsen	46.15	0.00
2. Standard Oil	46.15	0.00
8. Texas Instruments	7.69	0.00
8. USMC	7.69	0.00
8. Westinghouse	7.69	0.00

score (7.69), which reflects the fact that it lay on the shortest path 'between' USMC and the six other actors in the network connected with one another. Clearly, the number of ties a firm has with others influences a firm's betweenness score, but as it is shown below, some firms' betweenness centrality scores were actually higher than their degree centrality scores. Table 6.4 also indicates that Shockley and Sperry have dropped off the top 10 centrality list, being replaced by Texas Instruments, USMC and Westinghouse.

By 1970 SV's semiconductor community had grown to 35 firms, a sharp increase from the 14 present in 1965. As Graph 6.10 indicates, the network is growing both in size and complexity. As the number of actors in the network increases, the structure that will shape SV's semiconductor community in later periods begins to form. Other semiconductor firms have clearly been active in the founding of new firms, but still lying at the centre is Fairchild along with the six other firms it joined with to found Rheem.

As Table 6.5 indicates, Fairchild's normalised degree centrality is more than twice that of any other actor in the network while its betweenness centrality is almost three times that of the only other actor in the network

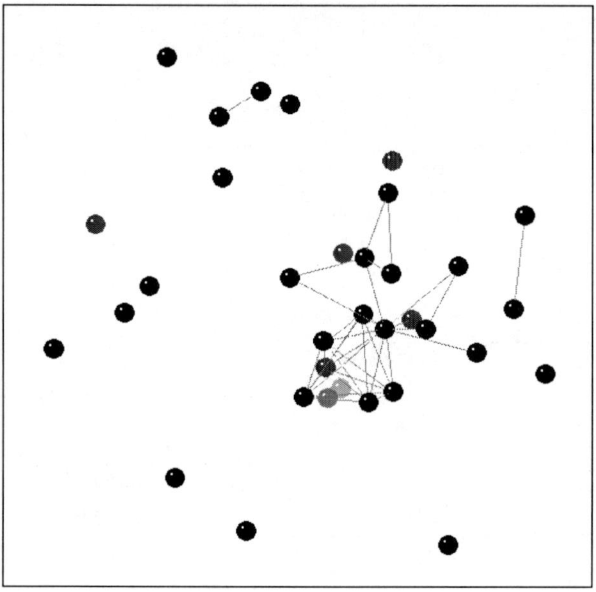

Graph 6.10 SEMI semiconductor community, 1947–70.

Table 6.5 Centrality ranking of SV's semiconductor firms, 1947–70 ($N = 35$)

	Degree centrality	*Betweenness centrality*
1. Fairchild Semiconductor	35.29	12.30
2. Bell Telephone Laboratory	17.65	0.00
2. General Atomic	17.65	0.00
2. General Transistor	17.65	0.00
2. Hughes	17.65	0.00
2. Morris-Knudsen	17.65	0.00
2. Standard Oil	17.65	0.00
8. Hewlett-Packard	11.77	4.28
9. General Instrument	5.88	0.00
9. IBM	5.88	0.00
9. Intersil	5.88	0.00
9. Mellonics	5.88	0.00
9. Signetics	5.88	0.00

that at this point in time registered a measure of betweenness centrality (HP). HP's emergence in 1970 as a central actor in the community is worth highlighting. By 1970 former HP employees had been involved in the co-founding of two new semiconductor firms: Qualidyne and Signetics Memory Systems. One former HP employee joined with two from Fairchild and one from Intersil to co-found Qualidyne, while another joined with two from IBM and two from Signetics (which was co-founded by six employees from Fairchild) to co-found Signetics Memory Systems.

Saxenian (1996) has also noted that HP has played a central role in the emergence of SV's unique entrepreneurial culture. She argues that in SV there exists a tradition of openness and informal exchange of information at social gatherings, trade association meetings and community conferences. People who gather at these events discuss the latest technological subjects, and she traces this tradition of informal exchange in part to David Packard's and William Hewlett's willingness to extend their personal assistance to other firms and entrepreneurs in the area, even to those who had been their direct competitors since the 1950s.

As Graph 6.11 illustrates, by 1975 the network structure of the community had begun to take a shape that resembles the most recent period in our analysis. Nineteen new semiconductor firms were founded in SV during the five-year period from 1970 to 1975, bringing the community's total to 54. What is clear from the graph, however, is that lying at the centre of this network was a handful of firms that seem to have been more active in the founding of new firms than others in the network. Fairchild once again lies at the centre of the graph, and its centrality is verified by its high degree and betweenness centrality scores (see Table 6.6), but other actors such as Intersil and HP are beginning to emerge as influential players as well. As Table 6.6 indicates in terms of their degree centrality HP and Intersil ranked behind only Fairchild, Bell Labs and Hughes, while in terms of their betweenness centrality HP ranked behind only Fairchild, while Intersil ranked behind only Fairchild, HP and American Microsystems.

By 1980, SV's semiconductor community's was beginning to take shape (see Graph 6.12). By then there were 63 actors in the network, and while Fairchild still lay at its centre, the central subgroup appears to have grown and become increasingly interconnected. Furthermore, there were fewer isolates in 1980 than in 1975. While in 1975 19 out of the 54 actors (35.19 per cent) in the network were isolates, in 1980 only 17 out of the 63 actors were (26.98 per cent), indicating that more firms and founders joined forces to co-found new firms. In 1980 Fairchild continued to be the most central actor in the community by both measures of centrality, while HP, Intersil and American Microsystems retained the high levels of centrality they began to demonstrate in 1975 (see Table 6.7).

In fact, the betweenness measures of centrality indicate that Fairchild, HP, Intersil and American Microsystems were the most central actors in the community. The final Graph 6.13 captures the state of the community in

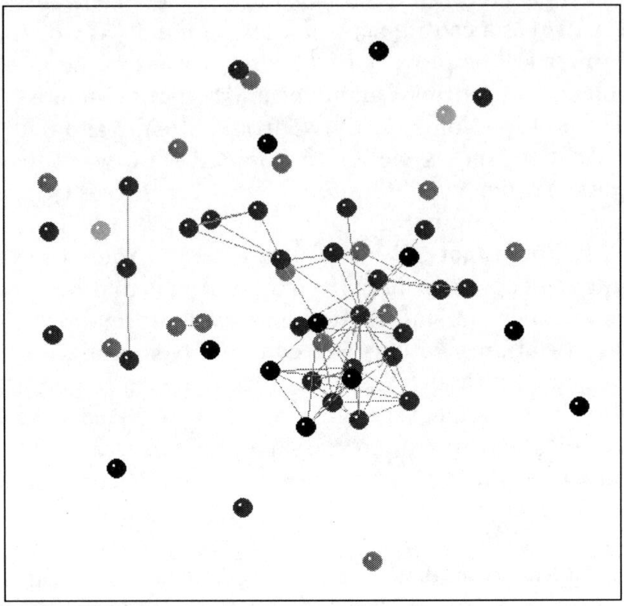

Graph 6.11 SEMI semiconductor community, 1947–75.

Table 6.6 Centrality ranking of SV's semiconductor firms, 1947–75 ($N = 54$)

	Degree centrality	*Betweenness centrality*
1. Fairchild Semiconductor	32.08	16.92
2. Bell Telephone Laboratory	18.88	1.39
2. Hughes	18.88	1.39
4. Hewlett-Packard	13.21	8.00
4. Intersil	13.21	2.65
6. American Microsystems	11.32	2.99
6. General Atomic	11.32	0.00
6. General Transistor	11.32	0.00
6. Morris-Knudsen	11.32	0.00
6. Standard Oil	11.32	0.00

1986. At that time the community consisted of 102 firms and was even more interconnected than it was in 1980. Of the 102 founding firms only 23 were isolates (22.55 per cent). The central actors largely remain the same, as Fairchild, HP and Intersil occupy the top of the centrality ranking (see Table 6.8). One remarkable development is the emergence of Intel as the second and third most central actor in degree and betweenness centrality respectively. It was not an influential actor at all as recently as 1975, but because of the influence of its founders, who are among the 'traitorous eight', it has become the most important firm among the 'Fairchildren' by 1986. Intel's

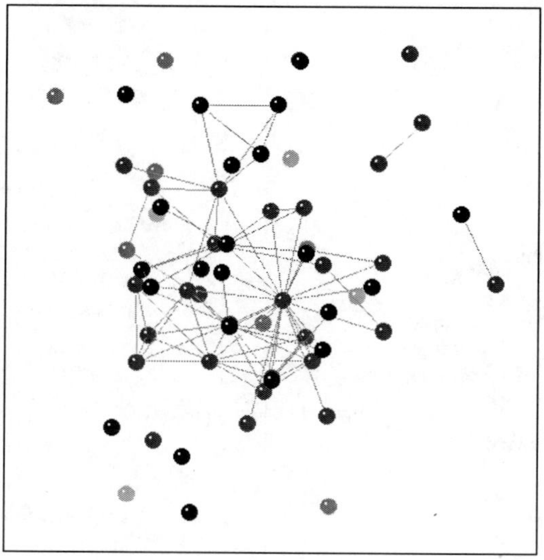

Graph 6.12 SEMI semiconductor community, 1947–80.

Table 6.7 Centrality ranking of SV's semiconductor firms, 1947–80 ($N = 63$)

	Degree centrality	*Betweenness centrality*
1. Fairchild Semiconductor	35.48	25.84
2. Bell Telephone Laboratory	16.13	1.41
2. Hughes	16.13	1.41
4. Intersil	14.52	7.25
5. Hewlett-Packard	12.90	13.21
6. American Microsystems	11.29	5.17
7. General Atomic	9.68	0.00
7. General Transistor	9.68	0.00
7. Morris-Knudsen	9.68	0.00
7. Standard Oil	9.68	0.00

centrality has most likely grown even further in the years following this period, but this hypothesis can only be confirmed by analysis of more recent data on the community.

This series of graphs and tables illustrate in some depth the evolution of SV's semiconductor community, from its origins in the late 1950s to the late 1980s. Starting with nine largely indistinguishable firms in 1960, the community had evolved into a complex network of ties among more than one hundred firms by 1986. In the process, Fairchild emerged as the most central actor in the community as early as the late 1960s, and maintained its status throughout the stages of its development. 'Fairchildren' firms such as Intel, Intersil and Signetics also score highly on centrality measures in subsequent stages. These findings underscore the central role of Fairchild and its

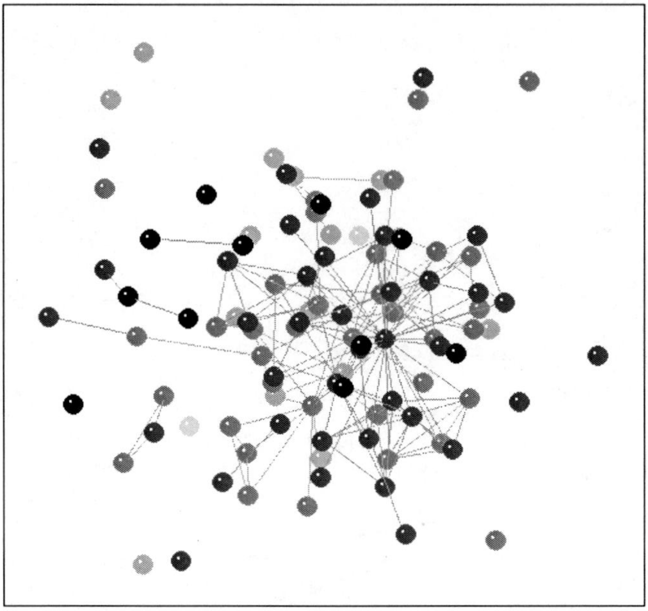

Graph 6.13 SEMI semiconductor community, 1947–86.

Table 6.8 Centrality ranking of SV's semiconductor firms, 1947–86 ($N = 102$)

	Degree centrality	*Betweenness centrality*
1. Fairchild Semiconductor	28.71	19.08
2. Intel	16.83	8.98
3. Hewlett-Packard	11.88	9.16
3. Intersil	11.88	7.84
5. Bell Telephone Laboratory	9.90	0.63
5. Hughes	9.90	0.63
7. National Semiconductor	8.91	4.90
8. American Microsystems	7.92	1.45
8. Signetics	7.92	3.02
10. Motorola	6.93	2.09

founders, the 'traitorous eight', in initiating and leading the semiconductor community, and help explain why its entrepreneurial and community-like culture diffused to the entire community, and even to cognisant technological communities such as venture capital. For example, Eugene Kleiner, one of Shockley's 'traitorous eight' founded, along with Thomas Perkins, the now top-ranked venture capital firm, Kleiner, Perkins, Caufield and Byers (see also Castilla 2003).

The founders of Fairchild believed in the culture of the democratic – horizontal – community, rather than the hierarchical workplace. This culture was passed on to Fairchild's spin-off firms and then to their spin-offs, thus spreading this horizontal culture of democratic community rather than hierarchical organisation in SV and its most successful firms. This culture diffused throughout the community and enabled collaboration across organisational and community boundaries, collective learning among large and small firms alike, high mobility of labour, informal exchange of information, and so on, all important factors in generating and sustaining the regional advantage (Saxenian 1996) SV has enjoyed through successive technological revolutions in the last few decades.

Figure 6.1 shows the cumulative number of firms in SV's semiconductor community from 1960 to 1986 (Assimakopoulos, Everton and Tsutsui 2003). It indicates that the development of this community follows the S-shaped curve proposed by Rogers (2003) in his model of the diffusion of innovations, and by others, such as Assimakopoulos (1997a, 2000) in his work on the emergence of new computer-based technological communities. Up to the late 1980s, it seems that at least three stages can be identified in the evolution of the semiconductor community in SV: innovators, early adopters and early majority. Shockley, Fairchild, Rheem and National Semiconductor seem to have initiated the innovators stage in the late 1950s. After 1965 it is possible to identify the formation of an early adopters category of firms subscribing to this community-like horizontal culture, originated at Fairchild and sustained by the creation of such significant spin-offs as Intel, leading to the creation of the critical mass of semiconductor firms in SV around the mid-1970s.

From the mid-1970s to the early 1980s the pace of company formation clearly slowed down (see Figure 6.1). This finding begs for further investigation into the decline and resurgence of the semiconductor community in SV in the late 1980s. During this period key founders such as Robert Noyce (Berlin 2005) played a leading role in setting up industry initiatives, such as the Semiconductor Industry Association (Van Stelle 2002), and even government policy, i.e. SEMATECH (Browning and Shetler 2000), fighting

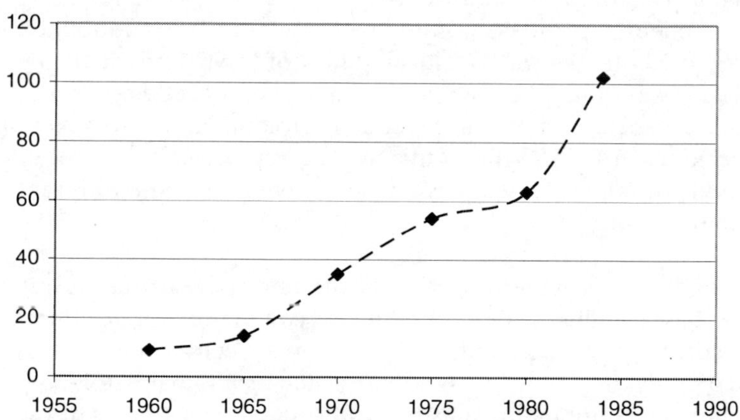

Figure 6.1 Cumulative number of firms in SV's semiconductor community, 1960–86.

fierce competition from abroad. The early majority of adopters seem to have emerged in the late 1980s, though the lack of data from the mid-1980s leaves the picture incomplete with respect to this category. It is worth also pointing out that in 1987 Fairchild itself was sold to National Semiconductor, which did not use its name for a decade or so. Fairchild Semiconductor re-emerged in 1997 through what has been called the first management buy-out in the semiconductor industry (Tuomi 2004).

The use of SNA, and in particular the combination of Ucinet, Pajek and Mage programs, enables us to carry out a multi-level analysis and explore the evolution of this new technological community and its particular industry structure with respect to both the networks of its founders and its firms founded. Moreover network analyses such as these can also uncover central actors that otherwise might have been missed. For example, the analysis above has repeatedly highlighted the centrality of Intersil Co. It was founded in 1967 and within its first decade it emerged as a major player in the founding of new firms (see Tables 6.5, 6.6; Graphs 6.4, 6.6). And by 1986 it ranked behind only Fairchild, Intel and HP in terms of centrality (see Table 6.8). Seldom, however, do histories of SV recognise the central role that Intersil has perhaps played in the formation of the semiconductor community. Saxenian never mentions Intersil in her account of SV, and in his historical account of Fairchild Semiconductor, Lecuyer (2000) mentions it only in passing, highlighting it as one of the many notable firms that Fairchild employees have helped to found over the years.

So far the discussion has focused on the semiconductor community itself, investigating its networks of both founders and firms founded. It has also highlighted the centrality of Fairchild Semiconductors and 'Fairchildren' spin-offs in the emergence of an entrepreneurial and horizontal community culture in SV. In parallel with industry it is also worth studying in more detail the role that universities in the region have played with respect to entrepreneurship and community development. For example, Fred Terman, the pioneering engineering dean at Stanford, played a leading role in fostering academic entrepreneurship and corporate involvement in ICT and related high-technology areas in the early critical stages of development of SV (Gillmor 2004). As was pointed out above, Terman encouraged Shockley, a Stanford graduate, to establish his transistor laboratory in Palo Alto next to the university campus in 1955. Without this single event perhaps the history of the semiconductor community and SV would have followed a different development path.

> Terman sought to strengthen the role of the university in supporting technology based industries by building a 'community of technical scholars' in the area around Stanford. In his words 'Such a community is composed of industries using highly sophisticated technologies, together with a strong university that is sensitive to the creative activities of the surrounding industry. This pattern appears to be the wave of the future'.

> In keeping with this program, Terman built the electrical engineering program at Stanford into one of the best in the country by recruiting promising engineering faculty and expanding its graduate programs. By 1950, Stanford was awarding as many doctorates in electrical engineering as MIT, despite its much smaller faculty.
>
> (Saxenian 1996: 22)

From a theoretical perspective, as was discussed in Section 3.2, recent non-linear innovation models focusing on the dynamics of science and research (Gibbons *et al.* 1994) and university–industry–government relations (Etzkowitz and Leydesdorff 2000) have shifted the emphasis to 'mode 2' knowledge production and the rise of the 'entrepreneurial university' (Clark 1998; Etzkowitz *et al.* 2000). Once more, social networks can provide insights into how these macro-innovation models link up to individual and organisational actions enabling innovation and entrepreneurship in SV through specifically professorial entrepreneurship and corporate involvement at the departments of electrical engineering and computer science at Stanford and UCB.

6.5 Professorial entrepreneurship at Stanford and UCB

The nature and structure of entrepreneurial activity by university professors has recently attracted considerable attention (Etzkowitz 2003; Shane 2004). SV universities, and in particular Stanford, have been renowned for fostering an entrepreneurial culture among their faculty for providing leadership and developing strong links with industry since the 1950s (Gibbons 2000). Kenney and Goe (2004) have studied the different levels of professorial entrepreneurship and corporate involvement in the EE and CS departments at UCB and Stanford. Building on earlier discussions about how economic action is embedded on social networks (Granovetter 1985) they used the concept of 'nested embeddedness' to explain the higher levels of professorial entrepreneurship at Stanford compared to UCB. Assimakopoulos and Kenney (2005) have expanded on these earlier findings, by visualising 'embeddedness' across multiple social relations connecting professors and firms in the region, and also exploring quantitatively the relational nature of leadership in professorial entrepreneurship at the two top research-led universities in SV.

Three sets of relations: 'founder' of a start-up firm, 'director' – member of a board of managing directors – and 'advisor' – member of a board of scientific or technical advisors – are analysed below as fundamental leadership dimensions for measuring entrepreneurship and corporate involvement for EE/CS professors at Stanford and UC Berkeley. More specifically, the relational data are derived by Kenney and Goe (2004) and can be seen in a series of two-mode 'professors by firm' graphs below. Based on these graphs, network measures are calculated with Ucinet 6 (Borgatti, Everett and Freeman 2002) for the networks of professors and firms involved. As expected, the analysis and findings suggest that Stanford professors of EE/CS have quantitatively demonstrated a higher level of leadership and corporate involvement than their colleagues at UCB with regard to starting and managing firms in the region. However, a surprising finding is that UCB professors demonstrate a higher level of leadership than their colleagues at Stanford with regard to the third dimension for sharing advice among themselves and with the firms in the region and beyond.

The initial design for this analysis included historical research into the extent of institutional support professorial entrepreneurship had received at the two universities, plus a two-stage survey of all EE/CS professors at Stanford and UCB (Kenney and Goe 2004). The original sample was the entire population of professors: 88 professors at Stanford and 92 professors at UCB were listed as members of the EE/CS departments in 2002. In the first stage an Internet search used the Google™ search engine to find any industry affiliations of these professors. The top 100 Google hits were analysed manually for each professor, although the departmental and company web sites were the ones that provided most useful relational data about each professor's affiliations. Based on this data collection exercise, in 2003 an email was sent to each professor asking them to verify and complete the results of the Internet survey. Responses were received from 44 Stanford and 43 UCB professors. It is important to note that the data is collected for the entire career of a professor so firms in the database need not have existed contemporaneously.

The initial study included the following affiliations: founder, president, CEO, vice-president, chair of board of directors, member of board of directors, chair of advisory board, member of advisory board, advisor, chief scientist, chief technical officer and 'miscellaneous' affiliations. These were re-classified into three types of relations: founders, directors and advisors. This simpler typology was thought to reflect corporate involvement and leadership more clearly and simply than the initial classification of industry affiliations, as university professors are assumed to have started, managed or advised firms in parallel with their academic careers. Stanford professors' affiliations with 11 mature and large firms (Cisco Systems, El Paso Natural Gas, France Telecom, Intel Microprocessor Research Lab, Hitachi America, Oracle, Lockheed Martin Corp., LSI Logic Research Lab, NCR, NTT DoCoMo USA Research Lab, Sperry Corp.) were removed from the original sample as it was thought that they did not reflect leadership and professorial entrepreneurship. The same applied to UCB professors' affiliations with the following four firms: Fuji Xerox Palo Alto Labs, Gerson Lehrman Group, NTT MCL and Siemens Tech to Business Corp.

Ucinet 6 (Borgatti, Everett and Freeman 2002) was used to compute network measures (see Appendix) such as centrality at the actor level, i.e. professor/firm, and at the university level, i.e. Stanford/UCB. Additionally for our networks they were calculated at the group level: cliques, density and the E-I index. According to Wasserman and Faust (1994: 251) the literature on cohesive subgroups has used various criteria to conceptualise the idea of a subgroup in a network. Three types of criteria are usually used to define cohesive subgroups. These are mutuality of ties, closeness or accessibility of subgroup members, and frequency of ties among subgroup members compared to non-members. Below a cohesive subgroup, i.e. a clique, is defined as a fully connected subgroup of a minimum of three actors, i.e. professors, firms.

The density of a network shows the proportion of connections that are actually present to the maximum possible. Given that a network consists of g actors then the maximum number of reciprocal links connecting these actors are $g*(g–1)/2$. If L links are actually present in a network then the density is calculated as $2L/g*(g–1)$ (Wasserman and Faust 1994: 101). Zero density means no links connecting the actors of a network, while density equal to 1 indicates that all the possible connections are present and the network is complete. Last, when we have a network, say depicting founder or advisor relations, with clear subgroups such as the Stanford and UCB groups of professors, then the 'external-internal' E-I index can be used to measure the extent of interactions within and across subgroups. In particular, with regard to UCB and Stanford professors of EE/CS, it is worth calculating the E-I index given the geographical proximity of the two universities for finding out whether the UCB and Stanford founder/director/advisor networks cross over and to what extent.

The E-I index was proposed by Krackhardt and Stern (1988: 127) as $EL–IL/EL+IL$ where EL = the number of external links, and IL = the number of internal links. As a result the E-I index can measure the degree of openness of each actor and subgroup or, put differently, the extent that two subgroups are inward- or outward-looking. The possible scores for the E-I index range from –1.0 to +1.0. When there are equal numbers of external and internal links then the E-I index is zero, but when there are more external links the index ranges from zero to +1 (+1 means there are only external links, thus an actor or subgroup is completely outward-looking). On the other hand, the E-I index is negative when there are more internal links. It equals –1 when there are only internal links and therefore an actor or subgroup is completely inward-looking.

The analysis of findings is organised in four sub-sections below. The first three sub-sections discuss findings according to the typology of founder, director and advisor relations. Each of these sub-sections discusses Stanford and UCB professorial entrepreneurship and leadership embedded in academic–industry networks, also presenting analysis of the patterns of relations among the firms these professors have founded, directed and advised in SV and beyond. The fourth part discusses findings across all three relations with regard to network density and the E-I index.

Leadership through founder networks

Kenney and Goe (2004: 703) point out that 'In total, UCB professors either solely founded or co-founded 36 firms, while Stanford faculty founded or co-founded 58 firms.' This difference was statistically significant; thus it was suggested that Stanford professors of EE/CS were more entrepreneurial than their UCB colleagues. Graphs 6.14 and 6.15 show all 36 Stanford and 26 UCB professors (circles) and the firms (boxes) they have founded or co-founded. As can be seen in Graph 6.14 there are three components where six Stanford professors have co-founded three firms: Teknowledge, Numerical Technologies and Sensys Instruments. Graph 6.15 is unconnected as no UCB professors have co-founded a firm.

Figure 6.2 shows the frequency of professors at the two universities founding firms. Overall Stanford professors have started more firms than their UCB colleagues in four out of five classes. But there is only one Stanford professor who has founded four firms, while there are two professors from UCB who have founded four. Two Stanford professors have founded three and five firms each respectively, while there was no UCB equivalent. On average, Stanford professors started 1.53 firms each, while UCB professors started 1.38 firms each. This seems a small difference, but as can be seen in Graphs 6.16 and 6.17 the Stanford-founded network is bigger and more closely knit than the UCB network. Note that a link between two firms exists in Graphs 6.16 and 6.17 when at least one professor has founded both firms. About 40 per cent (23/55) of the Stanford-founded firms are isolates in Graph 6.16, while about 55 per cent (20/36) of the UCB-founded firms are isolates in Graph 6.17.

Moreover the analysis of cohesive subgroups in Graphs 6.16 and 6.17 shows that the network of Stanford-founded firms has five cliques: 1 – Actel, Lightspeed, Numerical Technologies, PIXIM, Silicon Architects; 2 – Numerical Technologies, Excess Bandwidth, Integrated Systems, Spectrum Infotech Private; 3 – Design Power, IntelliCorp, Intelligenetics, Oxford Molecular, Teknowledge; 4 – CommerceNet, Mergent Systems, Teknowledge; 5 – FreeSpace Communications, Matrix Semiconductor, Xagros; while the network of UCB firms has only two cliques: 1 – Endobionics, Therafuse, Xactix, Verimetra; 2 – Cadence Design Systems, Synopsys, Accent and Comsilica.

Centrality analysis also reveals that the mean of degree centrality for Stanford-founded firms is 2.63 (standard deviation 3.19) while the mean of degree centrality of UCB firms is only a bit lower at 2.54 (standard deviation 3.42). Numerical Technologies and Teknowledge are the two most central firms in the Stanford-founded network with degree centrality scores of 12.96 and 11.11, and betweenness centrality scores of 0.84 and 0.56 respectively. On the other hand, Cadence, Synopsys, Accent, Comsilica, Endobionics, Therafuse, Xactix and Verimertra are the most central firms in the UCB-founded network, with equal degree centrality of 8.57 and zero betweenness

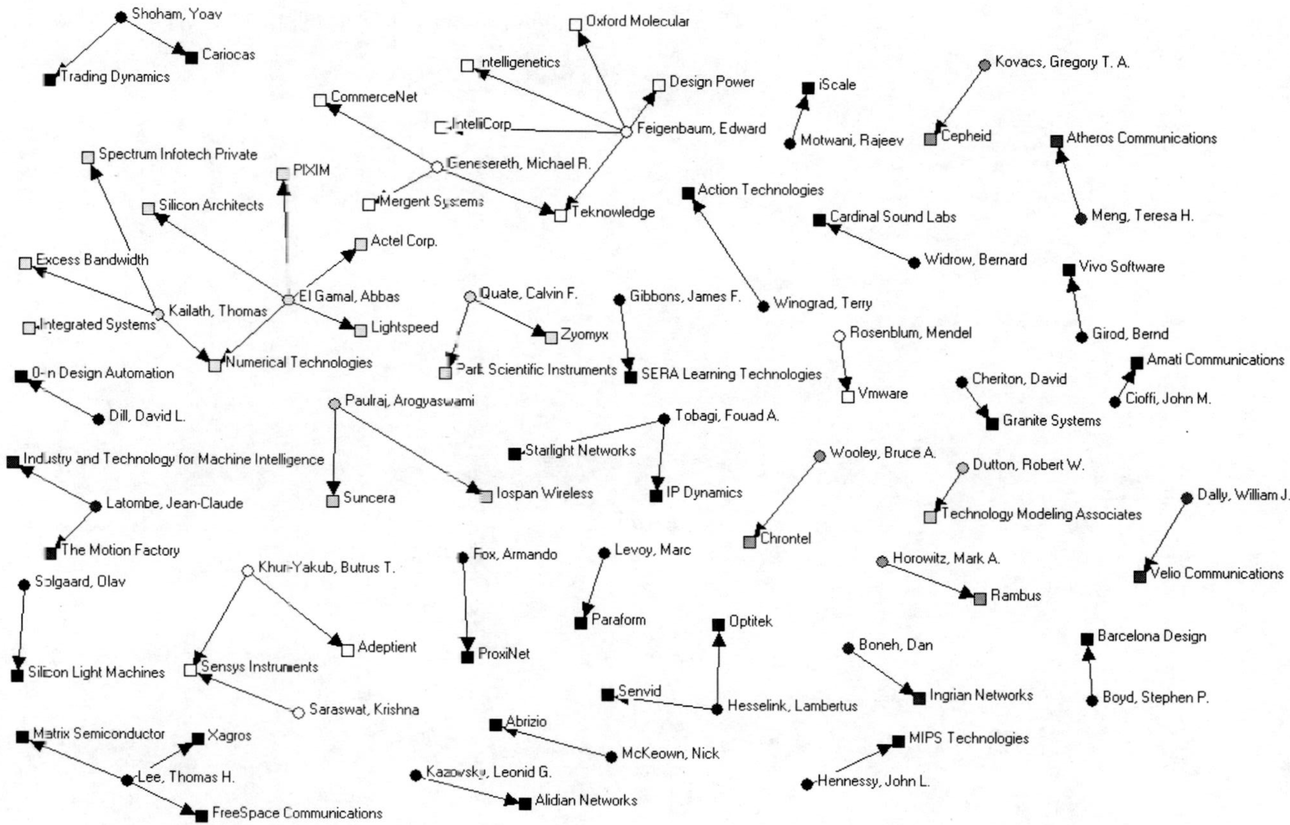

Graph 6.14 Stanford EE/CS professors by firms founded.

Graph 6.15 UCB EE/CS professors by firms founded.

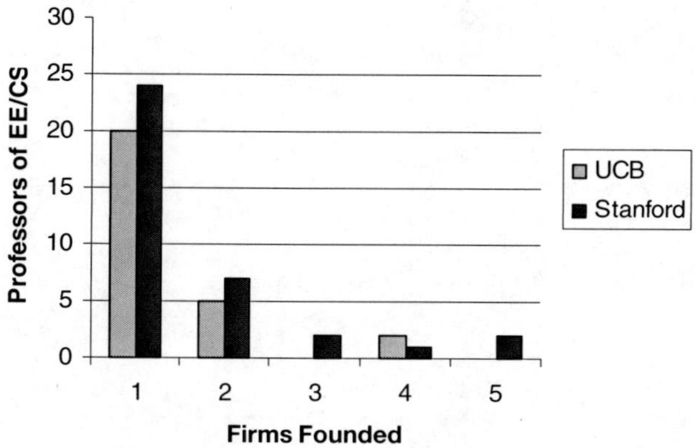

Figure 6.2 Frequency of founded firms by professors of EE/CS at UCB and Stanford.

centrality. As it was pointed out above, the founders' network of UCB firms is unconnected as no professors have joined forces to co-found a firm. As a result there is no betweenness centrality score for any UCB firm. On the other hand, Numerical Technologies and Teknowledge have more ties to the rest of the network of Stanford-founded firms (i.e. higher degree centrality) compared to their eight UCB equivalents. Both also stand at the crossroads of the cliques that have been co-founded by Stanford professors Thomas Kailath, El Gammal Abbas, Edward Feigenbaum and Michael R. Genesereth (i.e. score in terms of betweenness centrality), reflecting a stronger leadership from Stanford faculty, with a more cohesive and thus entrepreneurial network of ties among Stanford faculty and firms founded.

Leadership through director networks

Graphs 6.18 and 6.19 show all 22 Stanford and 16 UCB professors (circles) and the firms (boxes) they have directed or co-directed during their academic careers. From the outset it is worth noting that some professors did manage some of these firms full-time while they were on leave of absence, while in other cases they sat on a board of directors without directly managing the firm on a daily basis. As can be seen from Graph 6.18 there is only one component involving two Stanford professors/co-directors sitting on the board of directors of Numerical Technologies, a firm they also co-founded. Like Graph 6.15, Graph 6.19 is unconnected, as no UCB professors have co-directed any firms. Overall, compared to founder (and advisor, see below) relations far fewer EE/CS professors have been directors or managers of firms (about 10 help to manage one firm at both universities, while more than 20 have founded at least one firm at both universities), though Stanford professors still manage a quarter more firms as a total compared to

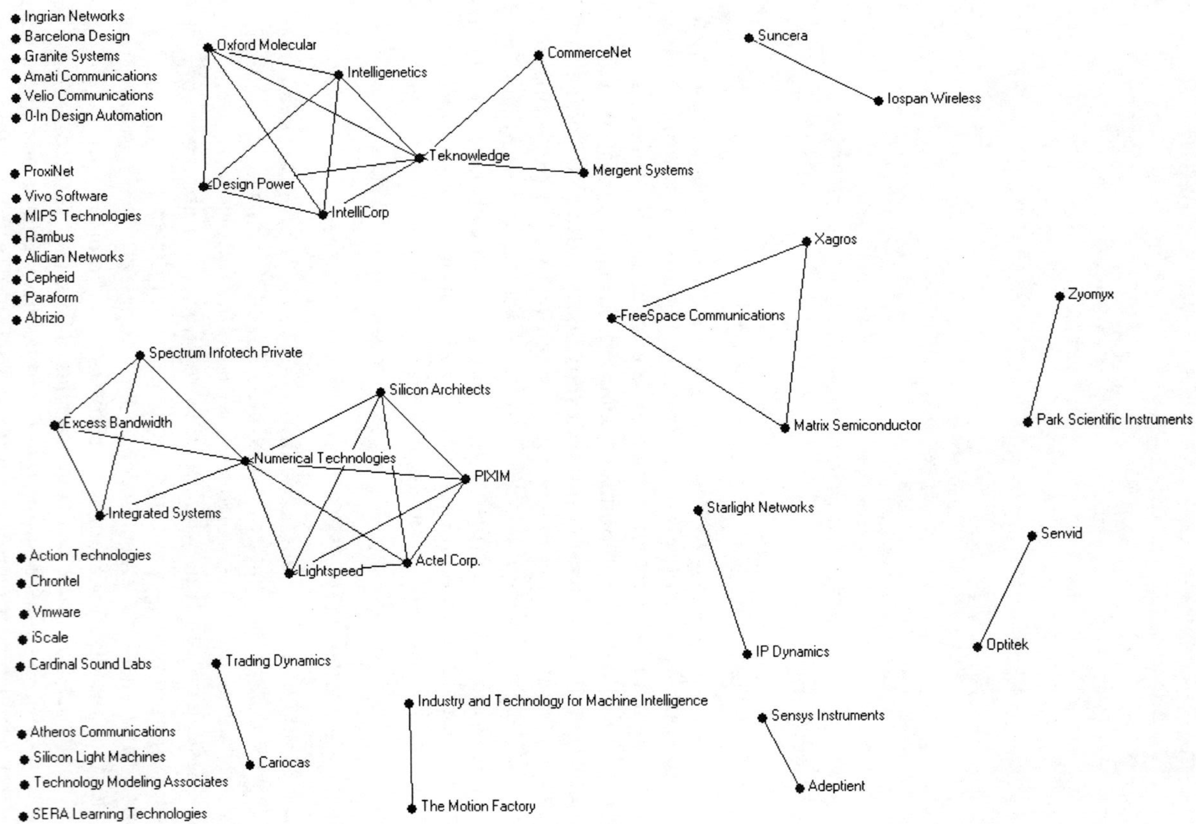

Graph 6.16 Stanford-founded firms.

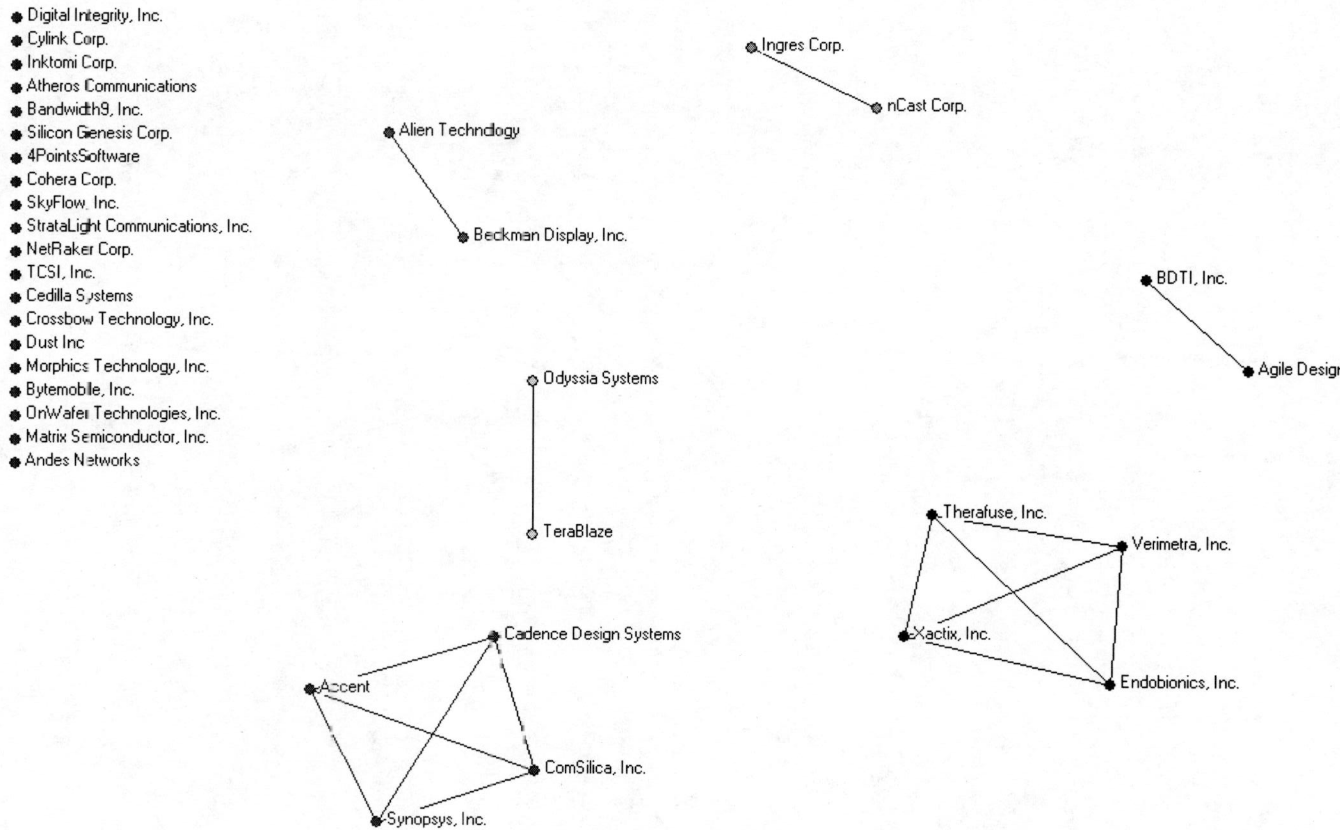

Graph 6.17 UCB-founded firms.

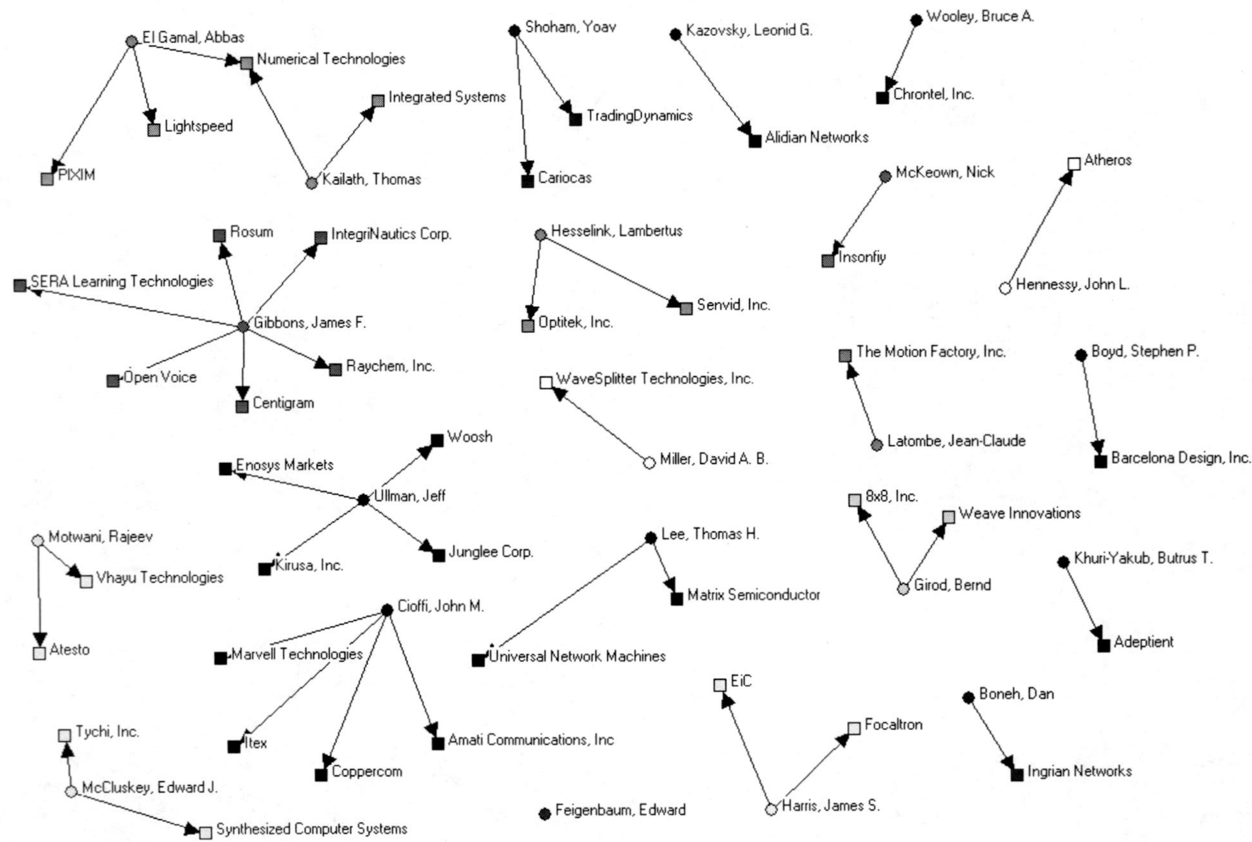

Graph 6.18 Stanford directors by firms.

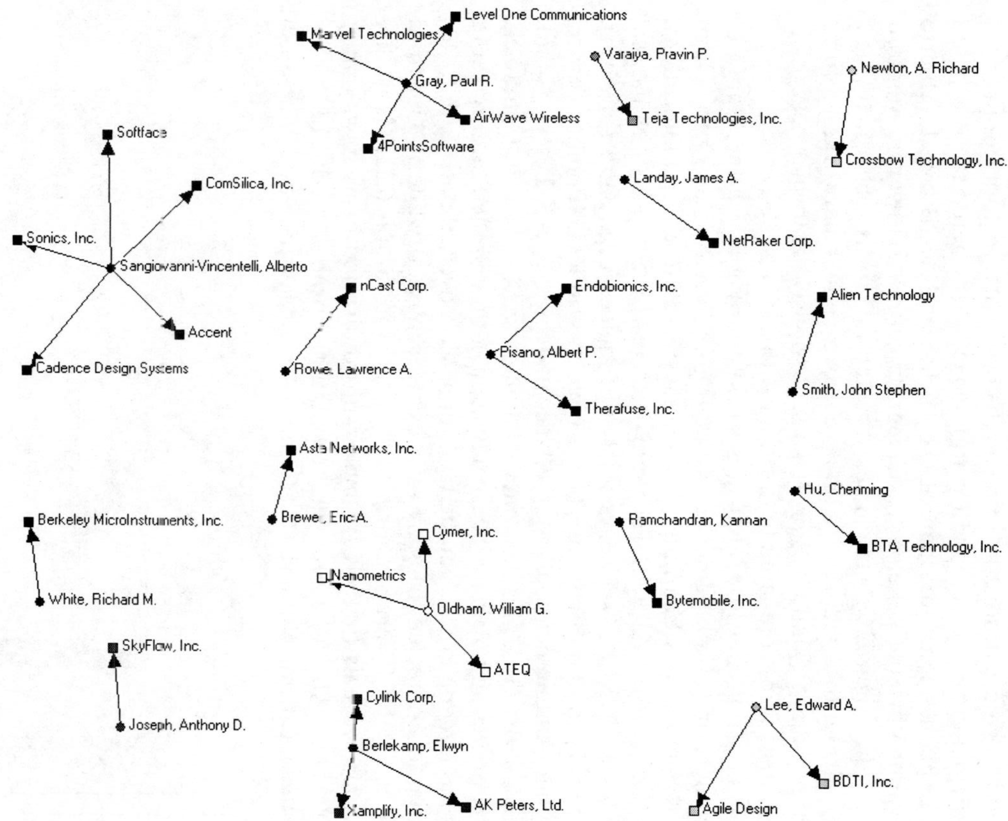

Graph 6.19 UCB directors by firms.

their UCB colleagues, demonstrating higher corporate involvement and leadership.

Figure 6.3 shows the frequency of directed firms by professors from the respective EE/CS departments. UCB has one more professor than Stanford sitting on the board of directors of two classes of firms (1 and 3). Stanford has six more professors than UCB sitting on the board of directors of two firms, and one more professor sitting on the board of directors of four firms. The same pattern is observed in the top two classes, where Stanford-based James Gibbons helps to manage six firms and UCB-based Alberto Sangiovanni-Vincentelli helps to manage five firms. However, Stanford professors, like their UCB colleagues, directed 1.81 firms each, on average. As a result, Graphs 6.20 and 6.21 reveal two similar networks of firms directed by Stanford and UCB professors. Obviously the former network is larger, with 41 firms compared to the latter with 28 firms.

The analysis of cohesive subgroups shows four cliques in the Stanford-directed network of companies (1 – Centigram, IntegriNautics Corp., Open Voice, Rosum, SERA Learning Technologies; 2 – Amati Communications, Coppercom, Itex, Marvell Technologies; 3 – Lightspeed, Numerical Technologies, PIXIM; 4 – Enosys Markets, Junglee Corp., Kirusa, Woosh), as many as in the UCB network (1 – AirWave Wireless, 4PointsSoftware, Level One Communications, Marvell Technologies; 2: AK Peters, Cylink Corp., Xamplify, 3 – ATEQ, Cymer, Nanometrics; 4 – Accent, ComSilica, Softface, Sonics and Cadence Design Systems), though the Stanford set of cliques has larger cliques than the UCB one. Moreover, about a fifth (9/40) of the Stanford-directed firms are isolates in Graph 6.20, while about a third (10/29) of the UCB-directed firms are isolates in Graph 6.21. Centrality analysis also reveals that UCB- and Stanford-directed firms have similar mean degree centrality scores, though Stanford firms scored a bit higher with only one firm, Numerical Technologies, today part of Synopsys, a

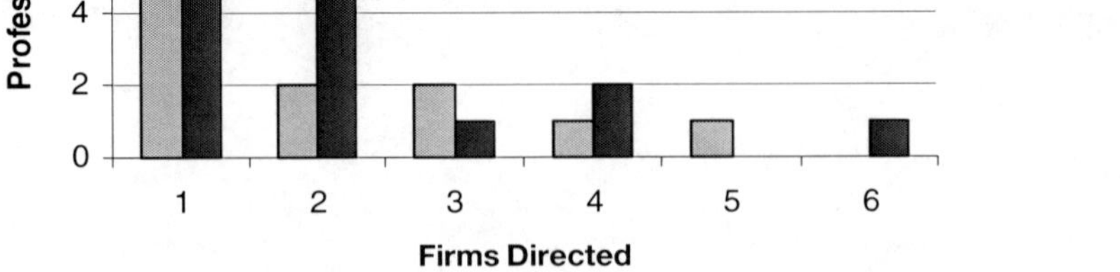

Figure 6.3 Frequency of directed firms by professors of EE/CS at UCB and Stanford.

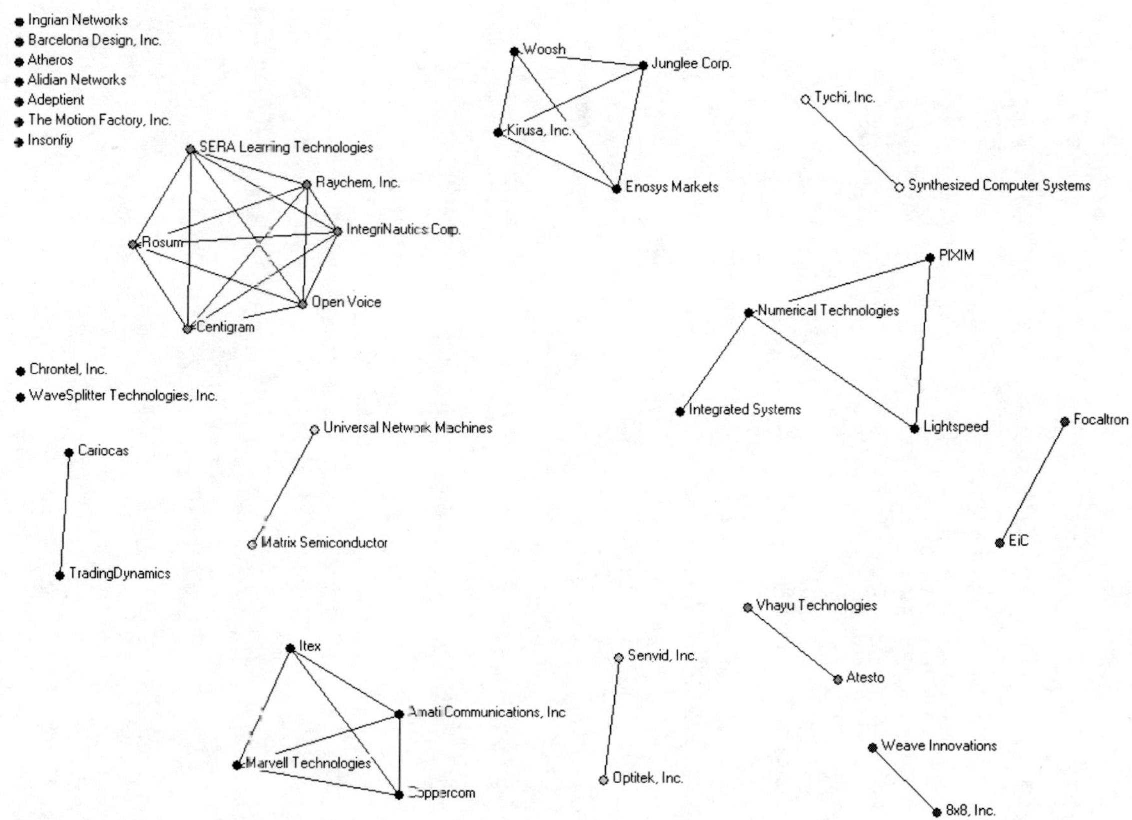

Graph 6.20 Stanford directors' firms.

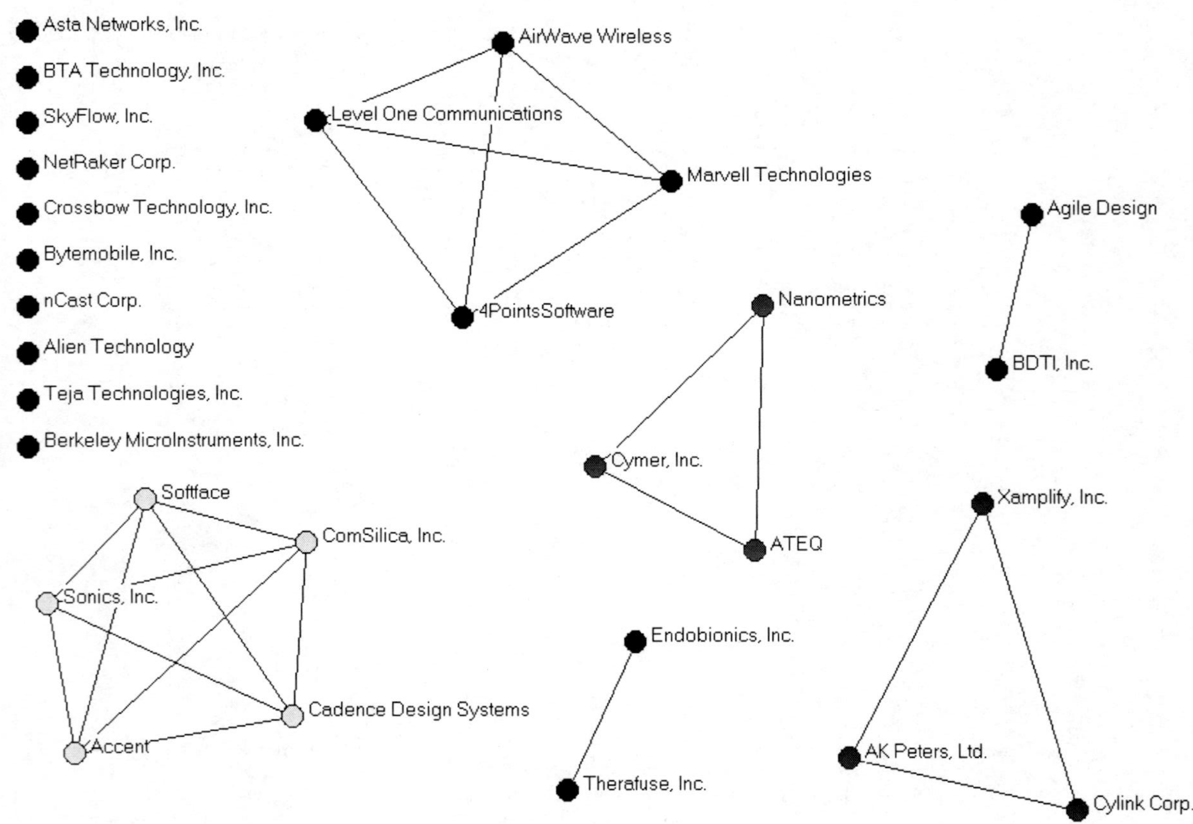

Graph 6.21 UCB directors' firms.

semiconductor software design firm co-directed by two Stanford EE/CS faculty, having a flow betweenness centrality score, indicating the willingness of Stanford faculty not only to co-found but also to co-direct SV start-ups.

Leadership through advisor networks

Graphs 6.22 and 6.23 show all 38 Stanford and 27 UCB professors (circles) and the firms (boxes) they have advised or co-advised. Moreover Graphs 6.24 and 6.25 show the networks of Stanford- and UCB-advised firms. Overall, as can be seen from these graphs, the network of advised firms related to Stanford is bigger (104 firms) compared to the UCB one (62 firms). A surprising finding, however, is that there are only two components in Graph 6.22 where three Stanford professors (Jeff Ullman, Rajeev Motwani and Terry Winograd) play advisor roles for Google, and two more (Bruce A. Wooley and Simon S. Wong) co-advise Atheros. On the other hand, Graph 6.23 shows that there are four much larger components where in the first three of them many more UCB professors co-advise 12 firms in total: Ardesta and Inktomi share three advisors each, and the remaining ones – SkyFlow, Crossbow Technology, Propel Software, Flarion Technologies, Simplex Solutions, Reshape, Tensilica, Chameleon Systems, Scale Eight and Visigenic Software – are co-advised by two UCB advisors each.

This finding is further analysed in Graphs 6.26 and 6.27 where the advisor networks among EE/CS professors at Stanford and UCB are shown. Note that in Graph 6.26 the vast majority, 87 per cent, of the Stanford professors are isolates, while in Graph 6.27 less than half (44 per cent) of the UCB professors are isolates. Furthermore, in Graph 6.26 there is only one clique (i.e. advising Google), while in Graph 6.27 there are two cliques: advising Ardesta (Albert P. Pisano, Kristofer Pister and Roger T. Howe) and Inktomi (David A. Patterson, Lawrence A. Rowe and David E. Culler). Unlike the founder and director relations above, UCB professors therefore seem to be quite active in playing the role of advisor or/and co-advisor compared to their colleagues at Stanford. On the other hand, Stanford professors are much more active in co-founding and co-directing firms as discussed above.

Figure 6.4 shows the frequency with which firms are advised by professors at the two universities. Stanford has more advisors in most of the classes, with notable cases Rajeev Motwani and James S. Harris with 11 advisor relations each and Jeff Ullman with nine. Twenty Stanford professors have advised at least one firm, compared to only ten from UCB. UCB professors only slightly outnumber their Stanford colleagues in the middle categories of three or four firms advised, with a notable case, Roger T. Howe, who had eight advisory relations. On average, Stanford professors gave advice to 2.6 firms each, while UCB professors gave advice to 2.3 firms each. However, isolate firms in Graphs 6.24 and 6.25 are at similar levels, a bit lower (11.5 per cent) for Stanford than for UCB-advised companies (13 per cent). Moreover when the patterns of advisory relations among firms is analysed

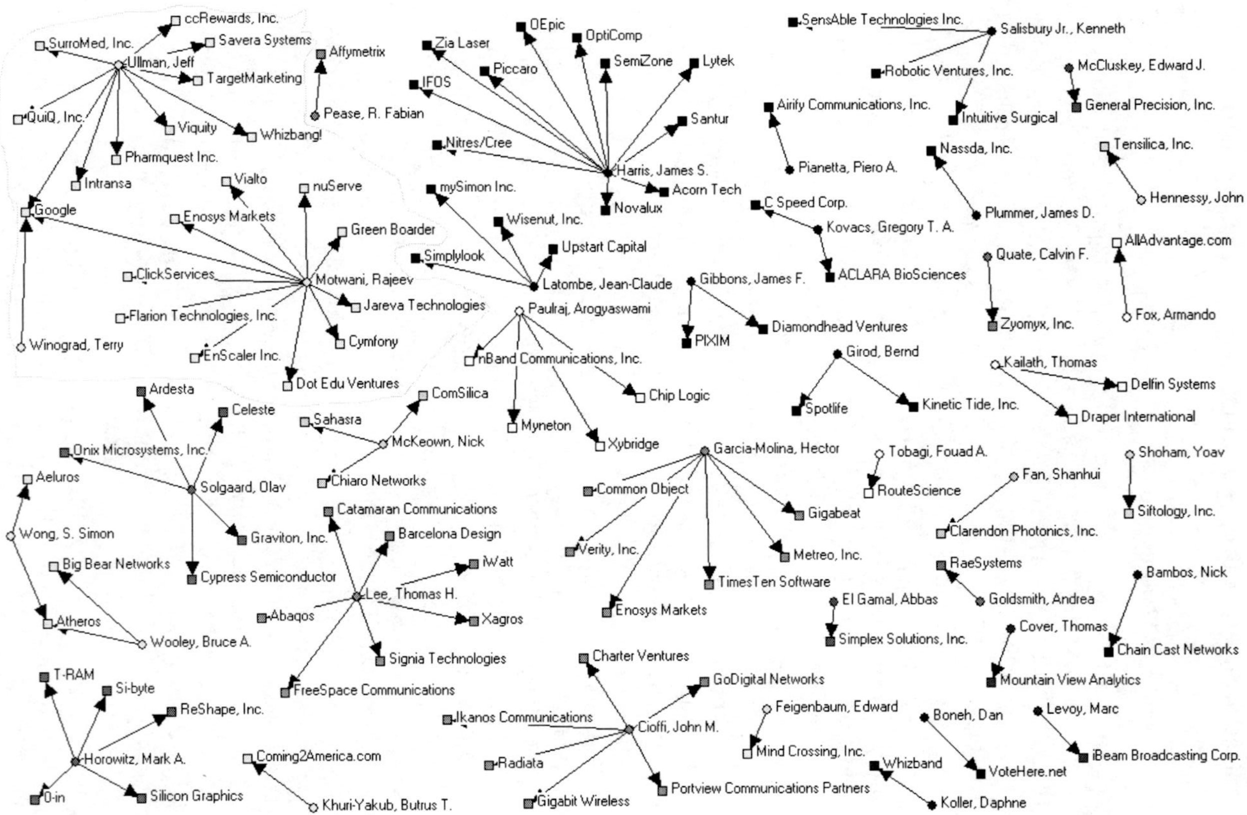

Graph 6.22 Stanford advisors by firms.

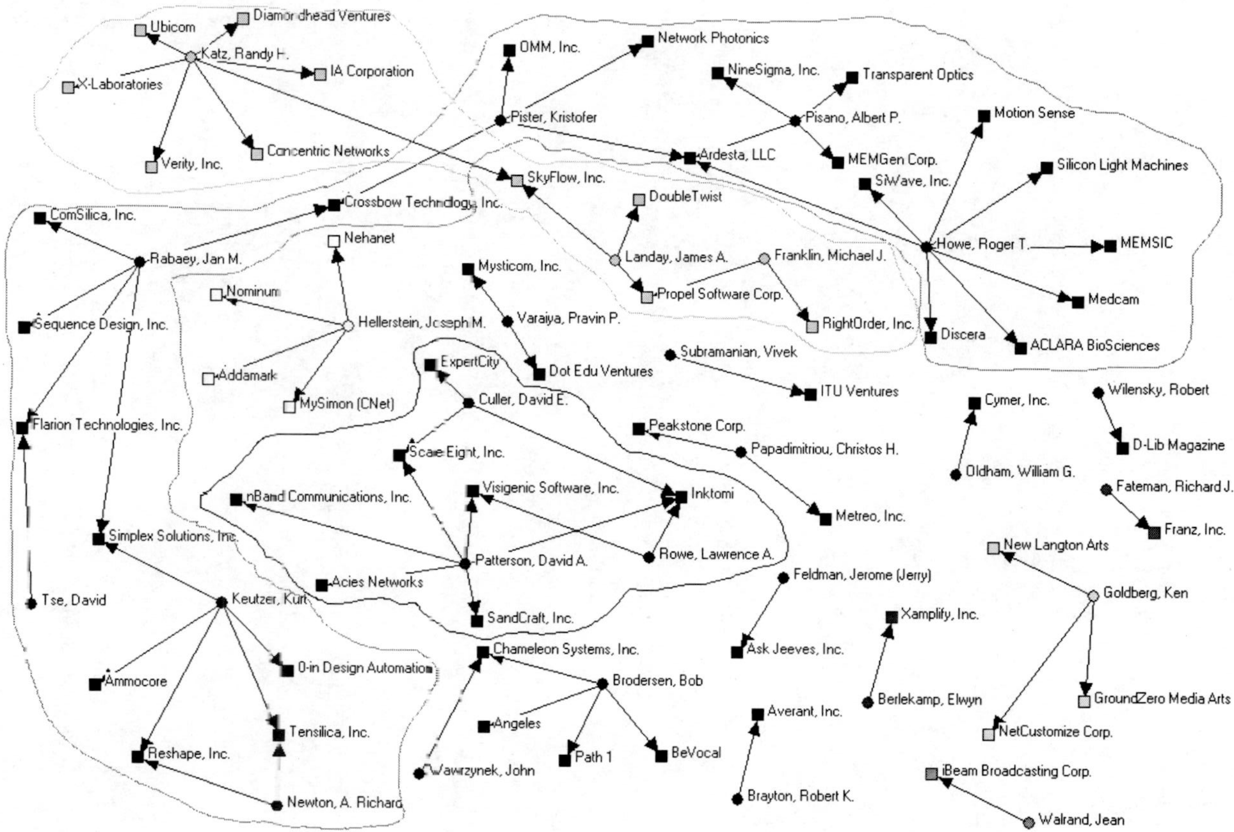

Graph 6.23 UCB advisors by firms.

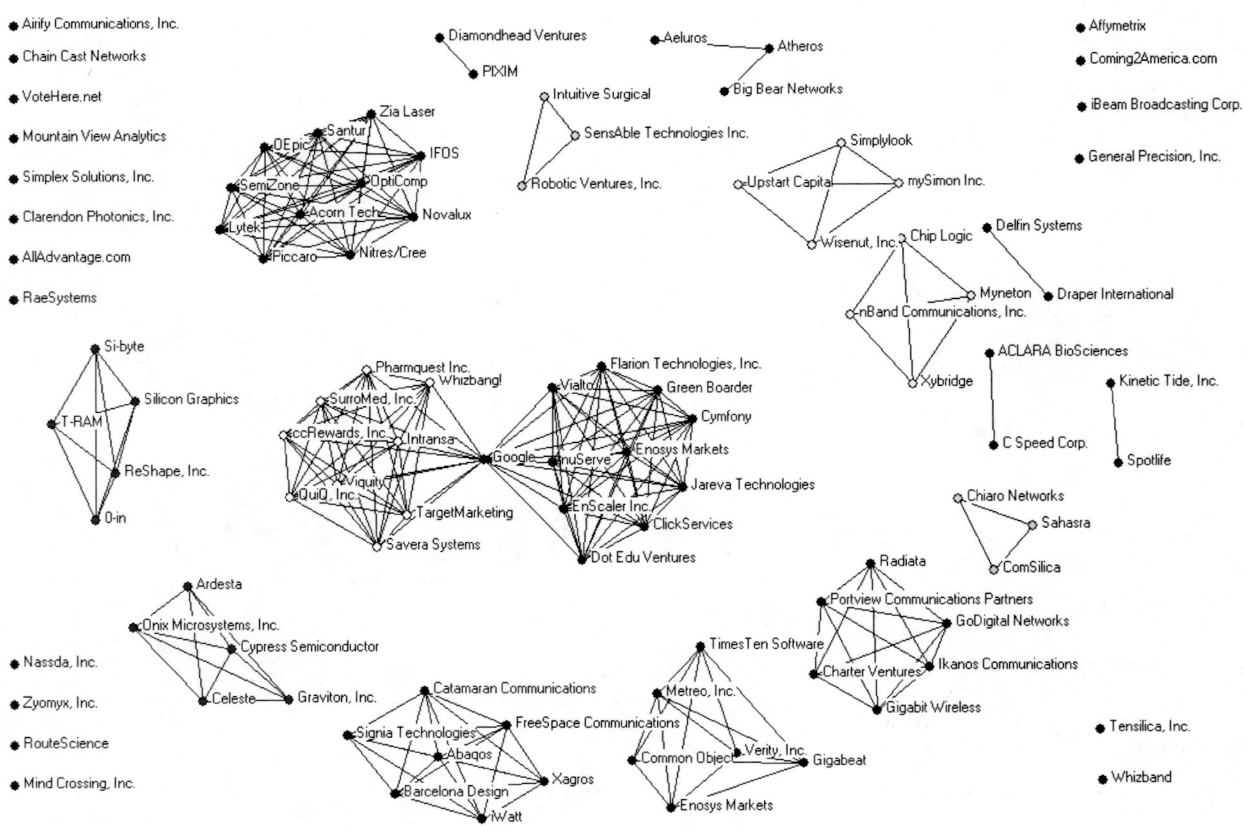

Graph 6.24 Stanford advisors' firms.

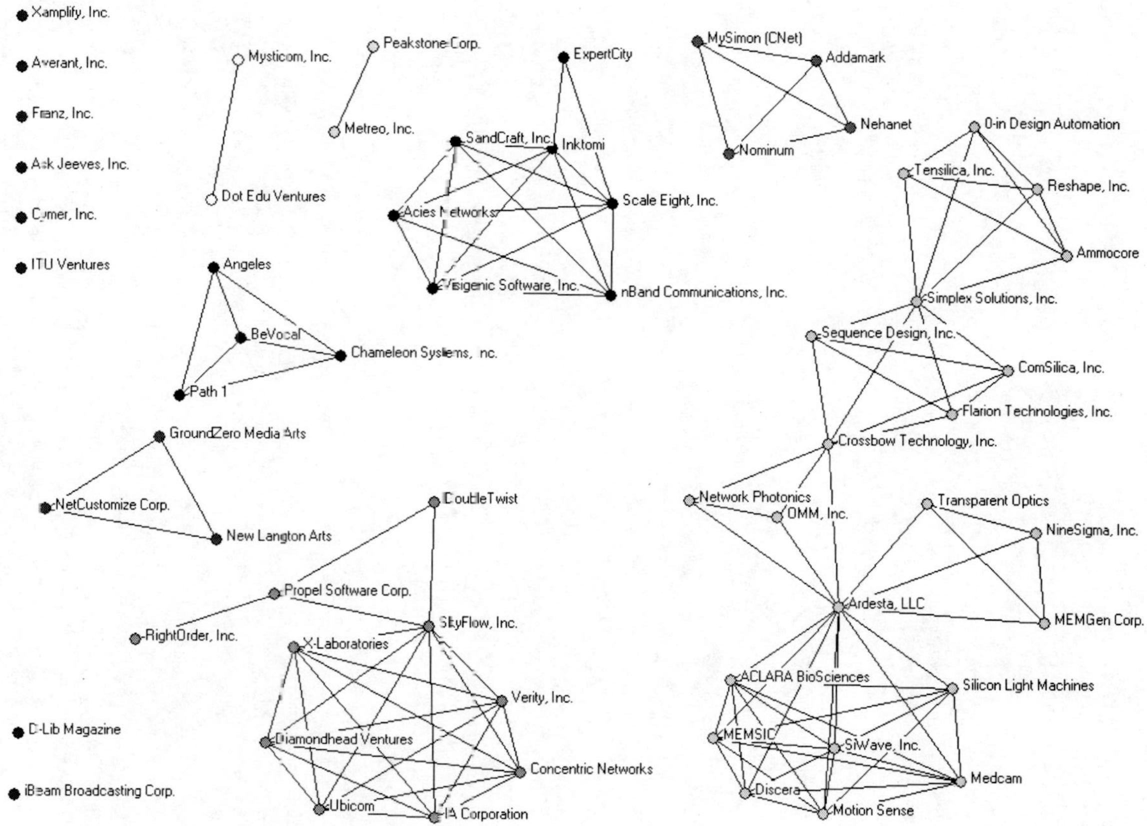

Graph 6.25 UCB advisors' firms.

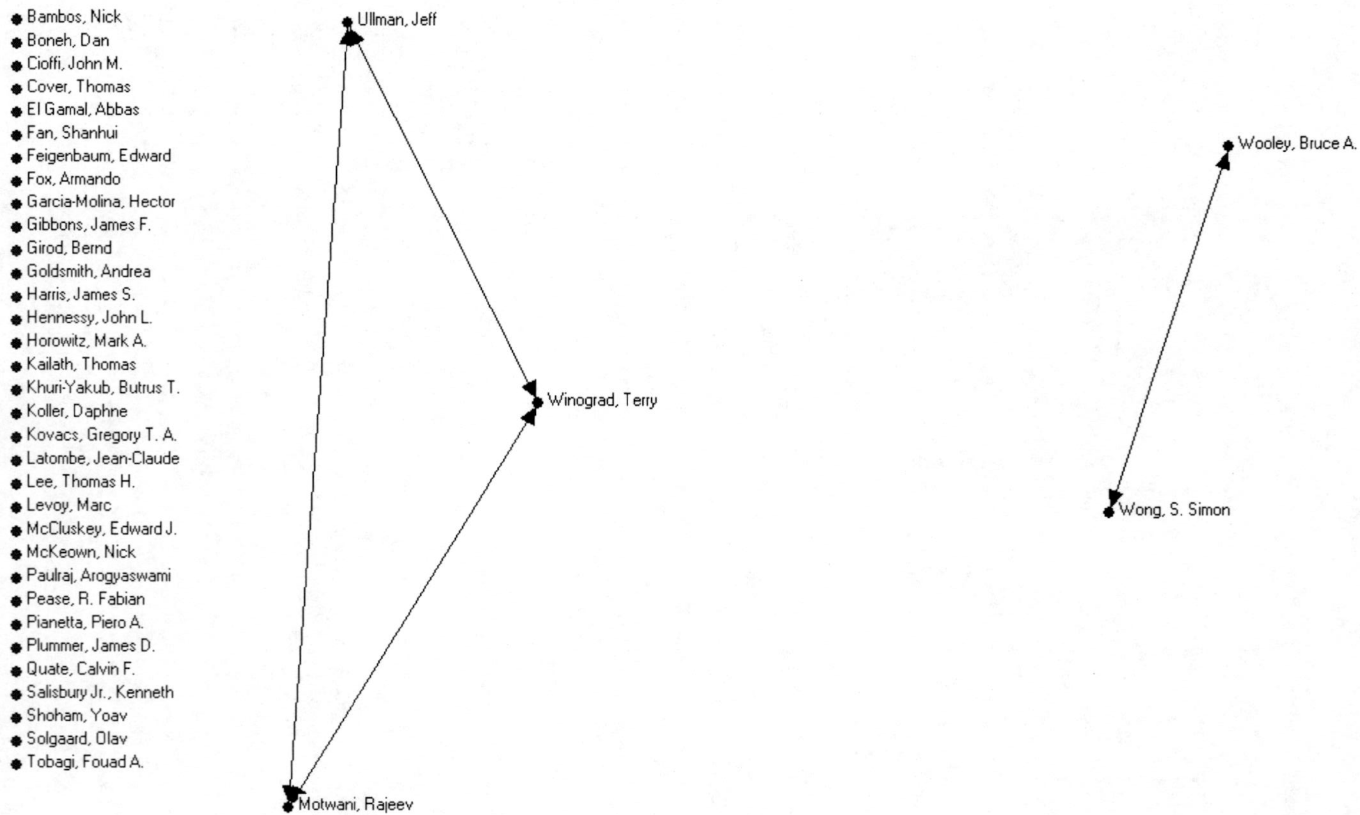

Graph 6.26 Stanford advisors.

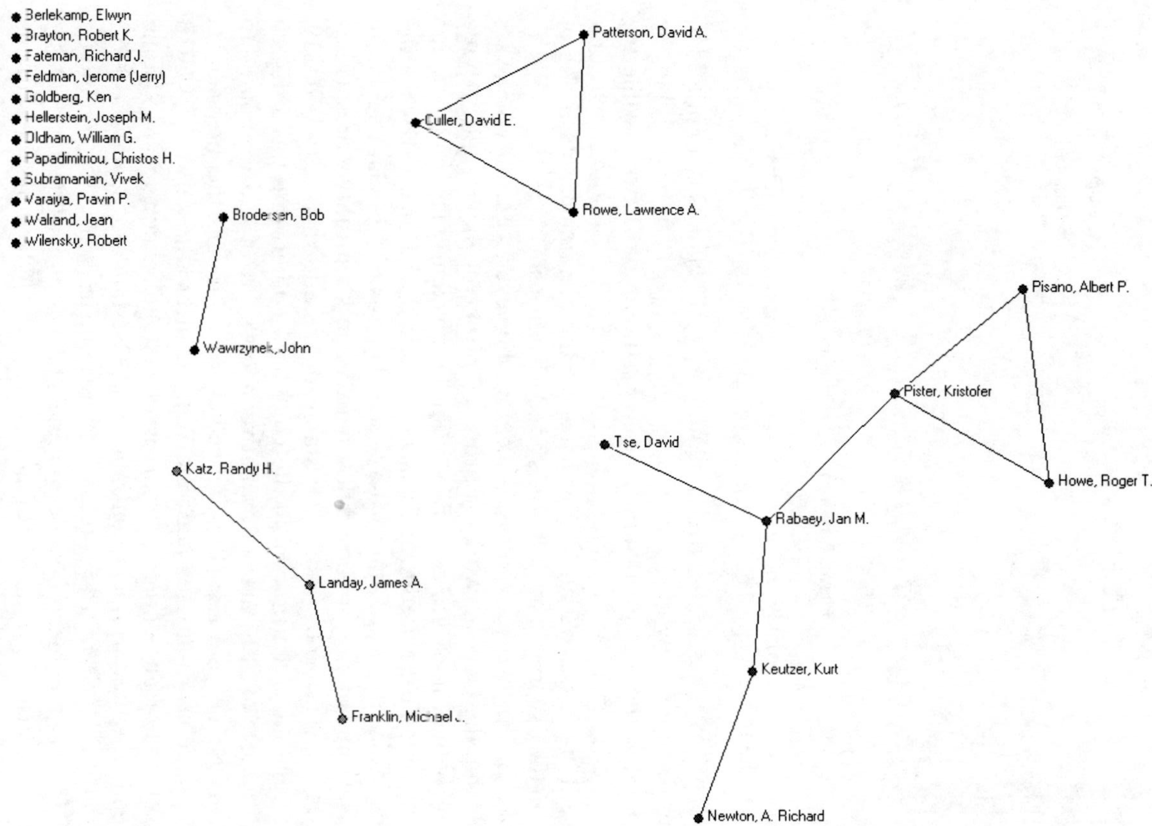

Graph 6.27 UCB advisors.

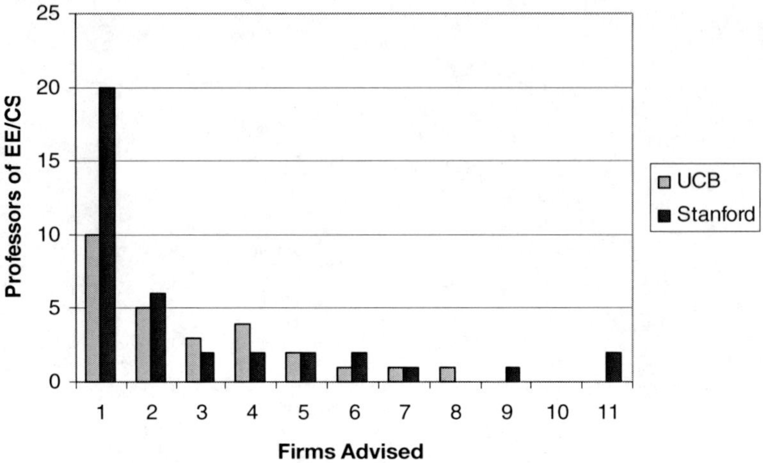

Figure 6.4 Frequency of advised firms by professors of EE/CS at UCB and Stanford.

with respect to cohesiveness the same number of 12 cliques is identified in both universities' networks, though Stanford's cliques are bigger than the UCB ones, as the former include up to a maximum of 11 firms, while the latter include up to a maximum of eight firms. This finding is further illustrated in Figures 6.5 and 6.6 which show the hierarchical clustering of cliques in the two university networks of advised firms.

Unlike in the two previous sub-sections, however, centrality analyses uncover that the UCB networks of firms and professors have higher degree and flow betweenness scores than Stanford equivalents. The mean degree centrality of UCB-advised firms is 6.29 (standard deviation 4.29), while the mean degree centrality of Stanford-advised firms is 4.67 (standard deviation 3.82). The same finding applies to flow betweenness centrality scores: UCB-advised firms – mean 0.50, std 1.34; Stanford advised firms – 0.07, std 0.19. Ardesta, the most central firm in the UCB-advised network, has a degree centrality score of 21.31 and a flow betweenness score of 7.72, compared to the 18.48 (degree) and 1.88 (flow betweenness) of Google, the most central firm in the Stanford-advised network. With regard to the network of UCB professors the mean degree centrality is 4.56 (standard deviation 4.91) and the mean flow betweenness centrality is 0.32 (standard deviation 0.82), compared to the Stanford professors' scores: mean degree 0.54 (standard deviation 1.48) and mean flow betweenness 0.05 (standard deviation 0.02). David A. Patterson and Jan M. Rabaey are the two UCB professors with the highest degree and flow betweenness centrality scores in the two universities' advisory networks.

The findings of the analysis with respect to the patterns of founder, director and advisor relations in both university networks are further discussed below using two network level-metrics: density and the E-I index.

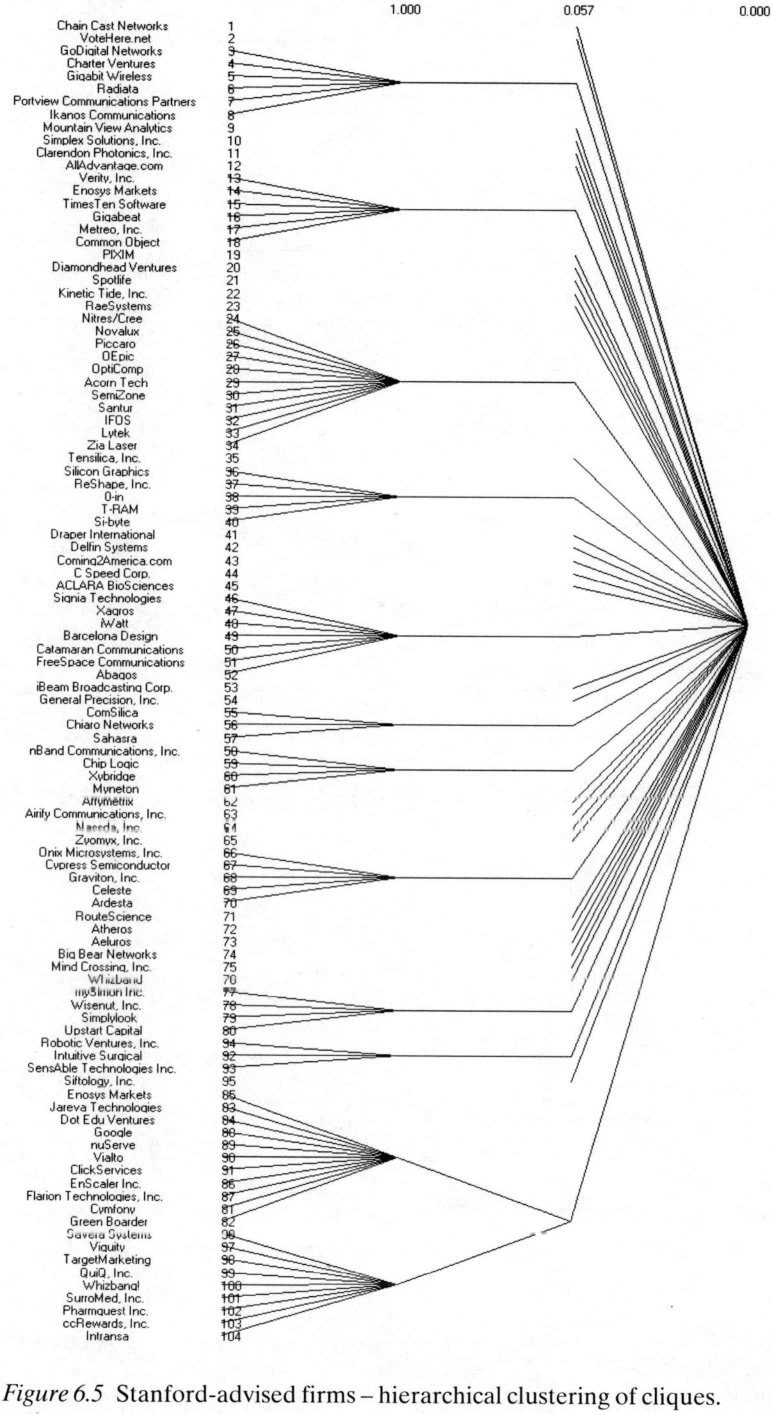

Figure 6.5 Stanford-advised firms – hierarchical clustering of cliques.

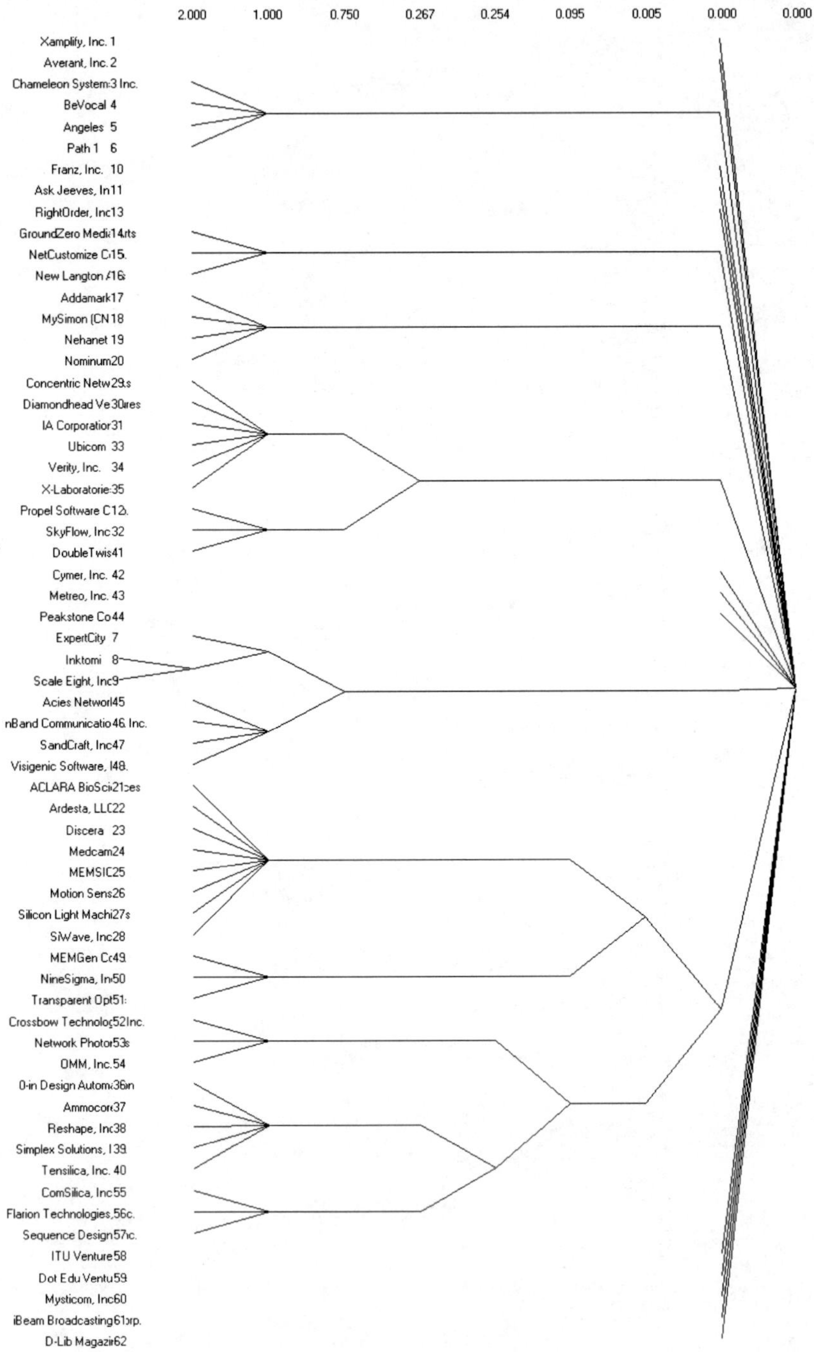

Figure 6.6 UCB-advised firms – hierarchical clustering of cliques.

Density and E-I index in professorial entrepreneurship networks

Table 6.9 shows the density of professors' networks at the two universities' EE/CS departments. In addition, Table 6.10 shows the density of firms' networks that have been founded, directed and advised by professors at the two universities' departments. As no UCB professors have co-founded and co-directed any firms the UCB networks of founder and director relations have zero density. Both Stanford-related networks of founders and directors and the networks of firms founded and directed by Stanford professors have higher densities than their equivalent UCB networks. However, both UCB networks of advisors and advised firms have higher densities compared to the Stanford ones suggesting that UCB professors demonstrate stronger leadership when it comes to co-advising and knowledge sharing among themselves and with firms in the region and beyond.

Table 6.11 shows the E-I index for EE/CS professors at both universities' founder/director/advisor networks. The results have to be interpreted with caution, however, as there are very few external and internal links with respect to the founder and director networks in particular. Note that a link for the E-I index, for example, in the founder network is external when two professors from both universities have co-founded a firm, while it is internal when two professors from the same university have joined forces to co-found a firm.

As was noted above, no UCB professors have joined forces to co-found any firm (i.e. $IL = 0$), but two UCB professors, Bob Brodersen and Vivek

Table 6.9 Density of founder, director and advisor networks at UCB and Stanford

Density	*Founder*	*Director*	*Advisor*
UCB	0	0	0.0456
Stanford	0.0048	0.0043	0.0054

Table 6.10 Density of founded, directed and advised firms' networks at UCB and Stanford

Density	*Founded firms*	*Directed firms*	*Advised firms*
UCB	0.0254	0.0529	0.0673
Stanford	0.0263	0.0599	0.0463

Table 6.11 E-I index of founder, director and advisor networks of EE/CS professors at UCB and Stanford

E-I index	*Founder*	*Director*	*Advisor*
UCB	+1.000	0	–0.209
Stanford	–0.500	–1.000	+0.172

Subramanian, have respectively co-founded Atheros Communications and Matrix Semiconductor with their Stanford colleagues Teresa Ming and Thomas Lee (i.e. $EL = 4$). As a result the E-I index for the founder network of UCB is +1 (4–0 / 4+0), indicating that UCB is completely outward-looking with regard to founding activity. On the other hand, as was noted above, six Stanford professors have co-founded Teknowledge, Numerical Technologies and Sensys Instruments (i.e. $IL = 6$). As a result, the E-I index for the founder network of Stanford is –0.5 (4–6 / 4+6) indicating that Stanford faculty are half inward-looking with regard to founding activity.

Table 6.11 also shows that the E-I index for the network of directors at UCB is zero, as UCB professors have neither internal nor external links with regard to managing/directing firms, while the Stanford network of directors is completely inward-linking (–1), as there are only two internal linkages connecting Thomas Kailath and Abbas El Gamal to Numerical Technologies. Overall, Stanford faculty therefore appear inward-looking with regard to both founding and directing relations, while UCB professors are outward-looking with regard to their limited founding activity, and they have not demonstrated any leadership with managing/directing any firms in the region. One of the reasons for the lack of UCB professors managing firms is the strict conflict-of-interest rules the University of California inflicts upon professors. These largely prohibit professors from accepting managerial positions in firms.

Last, Graph 6.28 shows all 65 UCB and Stanford professors (circles) and 149 advised firms (boxes); and Graph 6.29 shows only the 65 advisors divided into two categories: UCB (light grey) and Stanford (dark grey).

As it can be seen from Graph 6.29, there are seven cliques mainly populated by UCB professors – 1: Ullman, Jeff (S); Motwani, Rajeev (S); Winograd, Terry (S); 2: Motwani, Rajeev (S); Rabaey, Jan M. (UCB); Tse, David (UCB); 3: Horowitz, Mark A. (S); Keutzer, Kurt (UCB); Newton, A. Richard (UCB); 4: Keutzer, Kurt (UCB); Newton, A. Richard (UCB); Hennessy, John L. (S); 5: Keutzer, Kurt (UCB); Rabaey, Jan M. (UCB); El Gamal, Abbas (S); 6: Patterson, David A. (UCB); Culler, David E. (UCB); Rowe, Lawrence A. (UCB); 7: Howe, Roger T. (UCB); Solgaard, Olav (S); Pisano, Albert P. (UCB); Pister, Kristofer (UCB). This finding clearly indicates that UCB professors of EE/CS, though they are not allowed to found and direct firms with colleagues at either UCB and Stanford, they do openly create and share knowledge among themselves and with the firms in SV through advisory positions on scientific and technology boards. It is also worth noting that the most central professor in the network of Graph 6.29 is also based at UCB, Jan M Rabaey.

This analysis also indicates that there are a few subgroups that have emerged so far. This is not entirely surprising, because these start-up firms are often the product of a professor or graduate student's idea and often do not require large teams. Unlike the founder and director networks, however, the advisory network of UCB and Stanford faculty is much more

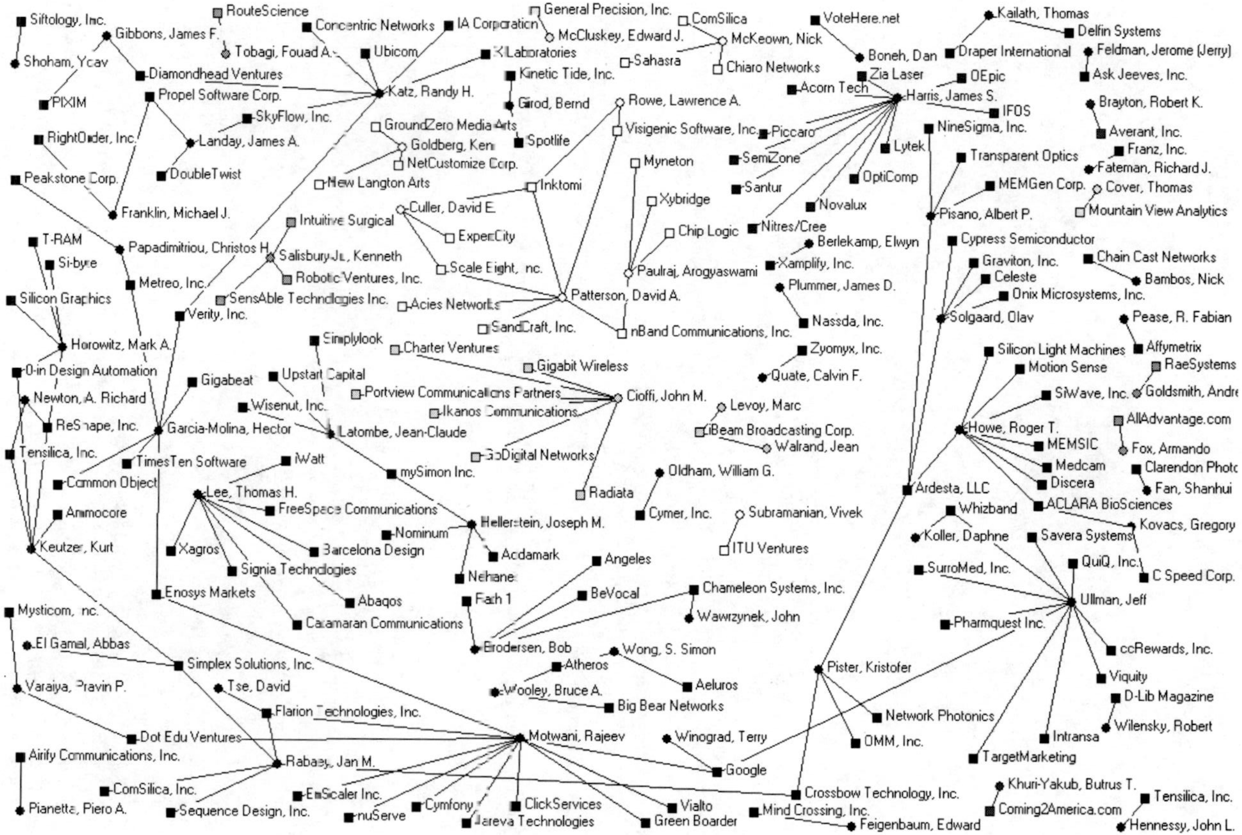

Graph 6.28 Stanford and UCB advisors by firms.

Graph 6.29 Stanford and UCB advisors.

overlapping, with UCB having 43 advisory links (*EL* = 17 and *IL* = 26), and Stanford having 29 advisory links (*EL* = 12 and *IL* = 17). Table 6.11 shows that the E-I index for the network of UCB advisors is –0.209, rather inward-looking, while the E-I index for the network of Stanford advisors is +0.172, rather outward-looking. All in all, Table 6.11 therefore indicates that Stanford faculty prefer to start and manage firms with colleagues from within the university, whereas when it comes to advising firms they share advice with UCB faculty. On the other hand, UCB faculty prefer to advise firms with colleagues from within the university, and in a few observed cases they also cooperate with Stanford faculty in starting firms within the region. Though they frequently advise firms, UCB faculty have been reluctant to co-direct firms with colleagues, most likely as a result of University of California rules that forbid professors from taking positions in the private sector.

The Stanford and UCB CS/EE departments are among the most entrepreneurial in SV. In parallel with their academic activities members of both departments have founded firms and served on boards of directors and advisory boards in SV and beyond. Network analysis indicates that, generally, Stanford professors have a greater number of entrepreneurial relationships than UCB in terms of co-founding and co-directing activities. Also, UCB professors are less likely to have corporate involvement, and when they do they are more frequently network isolates. Though professors from the two universities collaborate little on founding firms, there are a significant number of cases in which start-up firms have both UCB and Stanford professors as advisors. With respect to advisory relationships in particular, UCB professors demonstrate higher centrality and subsequent leadership and corporate involvement than their Stanford colleagues.

6.6 Summary

This chapter has explored the origins and early critical stages of development of the semiconductor community in SV. A multi-level network analysis highlighted the crucial role of a handful of pioneering individuals and firms that triggered and shaped this technological community and tradition of practice (Constant 1987) related to the production of ICs, which underpinned much of SV's success from the 1960s onwards, through successive generations of miniaturised digital technologies, leading up to the technological revolutions of the personal computers in the 1980s, and the Internet-related technologies of today.

Section 6.2 introduced the SEMI genealogy chart and told the story of how the first semiconductor firms were established in SV in the late 1950s. William Shockley was a star scientist and Nobel Prize winner who was attracted back to his home town to start working near Stanford University in 1955. His management style, however, failed to keep his firm and people on track for producing transistors for long. Out of Shockley Transistors, eight highly talented young scientists and engineers, the 'traitorous eight' started up Fairchild Semiconductor in 1957. The community-like (Wenger 1998) and entrepreneurial culture firstly initiated in Fairchild, coupled with state-of-the-art products, including the first IC patented in 1959 by Robert Noyce the head of its R&D department, set up a truly innovative organisational and business model that spread in SV through successive generations of dozens 'Fairchildren' spin-offs, including Intel in 1968, giving to SV its 'regional advantage' (Saxenian 1996).

Section 6.3 presented the inter-personal networks of the founders of semiconductor firms in SV, as well as the inter-organisational networks of firms that these founders came from, before analysing the network structure of SV's 'regional advantage' through the networks of founded semiconductor firms in Section 6.4. A series of graphs and tables highlighted the centrality of Fairchild Semiconductor and its founders, the 'traitorous eight', including such key inventors as Jean Hoerni, Robert Noyce and Jay Last, who socially constructed a revolution around a new technological community (Constant 1987) and network of practice (Brown and Duguid 2001). Graph 6.1 and Table 6.1 show the SEMI founders and that two-thirds of the top 20 most central individuals in this new technological community and network of practice were Fairchild founders and employees. The main component of this technological community was shown in Graphs 6.3 and 6.4, as well as the main component of its founders' previous firms in Graph 6.6. Table 6.2 also showed the top centrality scores of the SEMI founders' previous firms, unsurprisingly highlighting Fairchild, Intel and HP.

Section 6.4 focused on the networks of the semiconductor firms themselves, using six points in time covered by the SEMI genealogy chart in 1960, 1965, 1970, 1975, 1980 and 1986 (Assimakopoulos, Everton and Tsutsui 2003). The community started with a handful of firms that adhered to the new technological tradition of practice led by Fairchild, and as early as 1965 (see Graph 6.9 and Table 6.4) it became clear that Fairchild was its most central player spreading its community-like and entrepreneurial culture with an impressive array of spin-offs, such as Amelco in 1961, Intersil in 1967, and above all Intel in 1968 (see also Graphs 6.2 and 6.4). Fairchild continued its dominance in the semiconductor community for more than three decades, though after the 1970s HP, and after the 1980s Intel and Intersil, also occupied very central positions in a community that grew to 102 semiconductor firms in 1986. But the role of SV universities, in particular Stanford, was also of crucial importance from the early 1950s. Key visionaries such as Frederick Terman, the engineering dean at Stanford, played a central role in establishing a technological community of scholars within the EE faculty (Saxenian 1996; Gillmor 2004), and an industrial park next to the campus for enabling the emergence of an entrepreneurial university with strong corporate involvement.

The university–industry links were therefore explored in Section 6.5. Professorial entrepreneurship in the EE/CS departments at Stanford and UCB was studied along the three dimensions of founder, director and advisor relations reflecting leadership and corporate involvement in the two top research-led universities in SV (Assimakopoulos and Kenney 2005). It was suggested that the role of 'institutional embeddedness' directly affected academic entrepreneurship and the network structure of founder and director relations, as Stanford professors of EE/CS showed a higher level of leadership and corporate involvement compared to their colleagues at UCB. As a result of the University of California's strict conflict-of-interest rules, UCB professors did not take positions as directors in private firms and showed much less founding activity than their colleagues in Stanford. However, UCB professors demonstrated higher corporate involvement and leadership with respect to advisory positions in industry. All indexes at the individual, i.e. centrality, and group/network level, i.e. cohesion, density and E-I index, showed that UCB faculty actively shared knowledge and advice among themselves, with their colleagues in Stanford, and above all with the firms in SV and beyond.

7 Technological communities and networks

New frontiers for knowledge-intensive innovation

7.1 Introduction

Instead of starting from communities of practice and the nature of RTD work in ICT within organisational boundaries, this book started from the informal networks and technological communities which increasingly shape 'distributed' ICT innovations across institutional and geographical boundaries in cross-national, national and regional scales of analysis. The case studies of ten Esprit project networks and informal personal networks contributing information for innovation to ICT RTD, the origins and emergence of the critical mass of the Greek GIS community, and the SV semiconductor community, including academic entrepreneurship in UCB and Stanford put together network maps of people, organisations and relevant social groups which have initiated new technological traditions of practice and shaped a broad range of ICT RTD work at multiple levels of analysis and across different geographical, institutional and disciplinary settings. This 'social constructivist' view of RTD work extends current discussions on CoPs across organisational boundaries, also providing a heuristic way to study socio-technical networks and heterogeneous institutional and disciplinary interactions underlying the development of new technological communities and traditions of practice at cross-national, national and regional scales of analysis. It also accommodates an increasing number of technological traditions of practice, relevant social groups, and other actor networks, who produce and share knowledge among key players, such as CoPs and NoPs, within and across organisations, and thus attach meaning to new ICT.

Section 7.2 summarises the main conclusions of this research with regard to technological communities and networks. Section 7.3 discusses the theory-related implications with respect to the theoretical models of technological and scientific communities, COPs, NoPs and, more broadly, technological collaboration and networking. Section 7.4 discusses the policy and practice-related implications for networking in RTD work in Europe and beyond. Finally, Section 7.5 discusses the implications of this research for future research in the significance of technological communities and networks in ICT RTD and other knowledge-intensive areas for a competitive future.

7.2 Summary of main findings

The analysis and evaluation of findings in the previous chapters shows that technological communities, such as the semiconductor community in SV, are triggered by a handful of talented engineers and scientists such as the 'traitorous eight' who left Shockley and started Fairchild back in 1957. These eight men were closely linked to the scientific foundations of their technological traditions of practice, working with such star scientists as William Shockley, a Nobel laureate. They invented the 'planar process' and the industrial manufacturing of ICs and patented the first IC in 1959. Above all, they triggered a technological revolution and a new tradition of practice in semiconductor and IC RTD work in SV. As was discussed in Chapter 6, the contribution of Fairchild and its founders extends well beyond breakthrough technical innovation, in that these founders started up an innovative business and organisational model that enabled a new type of entrepreneurial and horizontal community organisation to emerge in SV, which diffused through dozens of Fairchildren spin-offs from the 1960s to the 1980s, most prominent among them Intel and Intersil Corporation. This tightly knit networked community of semiconductor founders became the foundation for SV's regional advantage in the 1980s and later, as it has propelled, with technical, managerial and financial support, successive generations of computer-based technological communities related to microprocessors in 1970s, personal computing in the 1980s, and to the Internet and mobile computing eras of the 1990s and early 2000s.

The role of knowledge-generating institutions, such as local research universities, in particular Stanford, from the beginning of the development of the semiconductor community in SV was also highlighted. Frederick Terman, the engineering dean at Stanford, built strong links with the semiconductor industry and nurtured a science park and technical community of scholars and entrepreneurs next to the university. A network analysis of professorial entrepreneurship at Stanford and UCB, the two leading research-led universities in SV, showed that institutional embeddedness (Granovetter 1985 and 2002; Kenney and Goe 2004) does matter with respect to the founder and director relations with their computer science and electrical engineering professors. Stanford faculty showed a higher degree of corporate involvement and leadership with respect to co-founding and co-directing relations than their colleagues at UCB. The University of California conflict-of-interest rules and regulations seem to have prevented UCB professors from taking positions as directors of SV start-up firms. However, UCB professors are more active than their Stanford colleagues in joining scientific and technology advisory boards and sharing advice among themselves and with their colleagues in Stanford and beyond.

The significance of 'glocalised' personal network communities for continuous technical innovation and industrial upgrading has also been increasingly recognised by commentators such as Saxenian (2001, 2005, 2006), who has recently introduced the term 'new Argonauts' for describing the reversal of what not long ago was thought to be 'brain drain' to 'brain circulation' between global technology regions. The latter is also certainly the case for the origins and emergence of the Greek GIS community, where the 'cosmopolites' who were educated and who socialised from the 1960s to the early 1980s in North America and Europe were the ones who initiated the new GIS tradition of practice in Greece, first in academia and subsequently in government agencies and private companies with funding provided by EC programmes. These academics have maintained their informal links and personal networks with the international research communities and have moved back and forth in Greece and abroad throughout the 1980s and 1990s, and across a broad range of institutional and disciplinary settings.

Furthermore, in Chapter 5 the heterogeneity (Law 1994) and essential complementarities of interests between the institutional, organisational and technological components (Orlikowski and Barley 2001) of the Greek GIS community were explored with socio-metric and SNA techniques, clearly illustrating the non-linear distributed nature of GIS innovations and capabilities found in 'mode 2' knowledge production (Gibbons *et al.* 1994) and the 'triple helix' of interactions among universities, government and industry (Etzkowitz and Leydesdorff 2000). Multiple memberships in the new GIS technological tradition of practice, as well as pre-existing technological communities in Greece such as spatial planners and surveying engineers related to geographic information handling and analysis, highlighted in particular the centrality of surveying engineering teams as the most prominent local relevant social group, and that its members occupied the most central positions in the emerging structure of the Greek GIS community in the late 1980s and early 1990s. During the past decade or so, these people, their teams, and supporting organisations have continued building a comparative advantage for the members of the surveying engineering technological tradition of practice in university laboratories (e.g. Cartography at the NTUA), government agencies (e.g. National Cadastre Organisation) and firms (e.g. Eratosthenis, Geomatics, Terra), as well as the main professional GIS association, HellasGIS, for the handling and analysis of geographic information.

Diffusion and use of GIS technology in Greece has therefore been shaped as a result of the adoption, implementation and continuous translation of the technology within a broad range of settings (Rogers 1993; Assimakopoulos 1997c). Technological brokering (Hargadon 2003), or multiple memberships in the new GIS-specific and pre-existing technological communities of teams that have adopted GIS, have propelled the development of the Greek GIS community along a certain technological path. In this unfolding trajectory the main emphasis was (and still is) on the development of topographic GIS databases and applications emphasising accuracy, rather than land use or socio-economic applications, in such crucial projects for national and regional development as the national cadastre, which has been co-funded by the EC and the Ministry of Environment, Planning and Public Works for the past decade or so.

Thus the underlying socio-technical structure and emerging professional culture (Constant 1987) related to the Greek GIS community have differentiated this GIS community from other national GIS communities in the Anglo-Saxon world (Assimakopoulos 2000). In the latter, physical and human geographers, not surveying engineers, were the most influential and prominent actors from the early critical days of socially constructing GIS innovations from both sides of the Atlantic (Couclelis 2004). In this respect, Greece falls within the same surveying engineering technological tradition that has influenced GIS diffusion in countries such as Germany, Austria and Italy. As will be discussed in more detail in the next section, the concept of technological community is therefore a useful theoretical construct for both exploring the essential complementarities of interest between institutional and technology components and creating a bridge between the macro-level analysis of GIS diffusion at country and/or cross-regional levels, and micro-level analysis at the team and personal egocentric network levels of analysis.

The significance of informal personal networks for innovation in ICT-related RTD work in Europe and beyond was highlighted in Chapter 4 through a comparative analysis of ten Esprit case studies. These project networks covered such diverse areas of ICT interest as hardware – for example, new asynchronous microprocessor architectures – software application areas ranging from manufacturing to banking to aerospace and defence, and softer issues, such as the building of worldwide consensus for trading multimedia property rights on the web. A main finding is that more than half of the nomination ties providing information valuable for innovation of the main UK Esprit contractors were both informal and personal in nature, external to both formal Esprit project networks funded by the EC and the organisational boundaries where the individuals who participated in this research came from. Despite a Euro-centric policy promoted by Brussels, more than three-quarters of the projects under study accommodated unacknowledged partners outside the European Union, located in places as far apart as Australia, Norway and Brazil. These unanticipated personal contacts provided connectedness and social capital (Burt 1992) to these informal innovation networks and NoPs, further underscoring the rise of networked individualism (Wellman 2002) in ICT RTD throughout Europe and worldwide.

Almost half (43 per cent) of the total number of ties identified in this research transcended the boundaries of the European Union and were external to the ten Esprit projects, throwing light on the global nature of these informal personal networks for ICT innovation. More than half of these external personal linkages were with people in the United States. Main UK contractors in Esprit valued these linkages with the most dynamic parts of the ICT world, in particular in California and on the US East Coast, solely for their information and knowledge content. The EC does not consider partners outside the European Union as equal partners for getting access to European public funding, and the only reason that can explain their 'unanticipated' existence is that people perceived them as very useful for ICT innovation. It seems that increasing emphasis on time-based and distributed ICT innovation and RTD work across the Atlantic, and across the oceans, will bring to light even more the value of personal networked communities such as the invisible colleges of academics, shifting the emphasis from formal organisations such as the RTD departments of firms and government agencies, and the exigencies of the latter for accountability related to public expenditure.

7.3 Implications for theory

A striking finding of the analysis above is that both the semiconductor community in SV and the Greek GIS community were locally and socially constructed by a handful of individuals, teams and organisations, that came from different institutional and disciplinary settings, but which shared a common interest in terms of instigating and building new technological communities and traditions of practice over the past five decades. Thomas Kuhn (1970) was the first to put forward the term 'paradigm shift' to explain how scientific revolutions were instigated by a handful of scientists who put forward a new paradigm, leading to the abandonment of an old one, and the parallel building up of a new invisible college (Crane 1972; Knorr-Cetina 1999). Edward Constant (1980 and 1987) built on the work of Kuhn and put forward the concept of technological community as the social locus of technological knowledge, innovation and practice. He argued that a community of technological practitioners, like a community of scientists, creates and follows a technological tradition of practice associated with the origins and evolution of a particular technology and linked innovations.

Based on the work of Constant, this research applied the concept of technological community and associated tradition of practice with particular emphasis on its knowledge (e.g. GIS software) and socio-cultural (e.g. social networks and structure of co-founder relations in SV) dimensions. It also extended it to the cognisant concepts of CoPs and NoPs (Wenger 1998; Brown and Duguid 2001; Wenger, McDermott and Snyder 2002), deploying a social network approach (Wasserman and Faust 1994; Cross, Borgatti and Parker 2002) to modelling the new semiconductor and GIS communities in SV and Greece. More specifically, the study of the origins and evolution of the SV semiconductor community triggered by the industrial production and patenting of the first IC in 1959 demonstrated that the technological revolution of semiconductor research in Fairchild and SV did not occur in a vacuum. The discussion in Chapter 6 went into some depth to highlight the origins and evolution of the community of semiconductor firms in SV, shifting the boundaries of theoretical discourse beyond any single genius inventor or technical breakthrough.

Key players for the emergence of the semiconductor community in SV, such as Fairchild and Fairchildren spin-offs like Intel, were started up and sustained by a team of talented scientists, engineers and entrepreneurs who were supported by and embedded within the knowledge and production networks of a specific geographical and institutional setting, near to Shockley Transistors and Stanford University, taking advantage of available funding from US government agencies, etc. In the process of critical mass formation for the SV semiconductor community a whole new technological community and tradition of practice evolved through the 1960s to the 1980s that was analysed through the networks of founders and founded firms, plus the academic entrepreneurship networks, presented in Chapter 6. This socio-technical ensemble embodied technological knowledge, innovation and social networks, laying the foundations for the regional advantage of SV (Saxenian 1996), and also paving the way for the emergence of the 'new Argonauts' (Saxenian 2006) and their 'glocalised' personal communities (Wellman 2002) in global technology regions in the early 2000s.

Like other computer-based technological innovations, GIS innovations in Greece have been neither adopted nor implemented in a vacuum from the mid-1980s onwards. They have been embedded within organisations and their CoPs, as well as in NoPs between various institutions, relevant social groups and the pre-existing technological communities which handle and analyse geographic information at national and international scales. Like other GIS diffusion studies (Campbell and Masser 1995; Harvey 2001), the research underlying Chapter 5 suggested that GIS should be conceptualised not according to a technological deterministic view, which puts the emphasis on computer equipment, software and digital data, but on a socio-technical view which shifts the emphasis of the discussion onto institutional, organisational and human issues (Barley 1990; Orlikowski and Barley 2001).

In contrast to the vast majority of GIS diffusion studies this research did not focus on specific organisations or the local government setting (Innes and Simpson 1993; Masser and Onsrud 1993; Masser, Campbell and Craglia 1996). It considered the whole pattern of socio technical linkages between academia, public sector agencies and private sector firms within the triple helix (Etzkowitz and Leydesdorff 2000) of university–industry–government relations at a national scale and as a result it explored the whole actor network of GIS teams and GIS linkages. These multi-stranded linkages highlight which teams and institutional and disciplinary groups have played a central role, or have been marginal, for leading GIS diffusion and building the new GIS community and technological tradition of practice within a whole country. As a result of this socio-technical view of GIS technology, a model was put forward in Chapter 5 showing the creation of the critical mass and diffusion and adoption of GIS innovations within Greece as a whole. This network model of a new GIS community can also be used for the analysis of GIS diffusion between different areas of the same country. Such a holistic network approach to the study of GIS diffusion is novel to the academic literature and provides both concepts and methods to study the patterns of linkages which highlight the interplay between the technology itself and the context within which it is embedded at national and cross-regional scales.

With these considerations in mind, this research shows that both GIS and semiconductor technologies can be viewed as new 'revolutionary' technological communities and traditions of practice, albeit of different socio-economic significance, consisting of a set of actors, individual founders, teams and organisations, and a set of internal and external linkages, highlighting the essential complementarities of interest between stakeholders, technology and community components. Each new tradition of practice linked to a specific technology was triggered and has evolved as the product of a heterogeneous array of knowledge-based economic, financial and social linkages between stakeholders which have gradually become interdependent at the national or regional scale because of essential complementarities of interests with respect to various components of GIS and semiconductor technologies.

New technological communities such as the Greek GIS community, however, are not autonomous entities within national boundaries, as the membership criteria for individuals, teams and organisations in such social systems are neither total nor singular. Unlike scientific communities, which require their members to abandon old paradigms of scientific practice, members of, for example, a new GIS community do not need to abandon old technological traditions of practice for the new one. Survey engineers, spatial planners and others who handle and analyse geographic information in Greece have tried to reinforce their positions in existing disciplinary and professional structures such as professional committees (Rosenkopf and Tushman 1998) and associations (Swan, Newell and Robertson 1999), while at the same time giving birth to a new GIS community which is creating new opportunities for a whole range of actor networks. This finding echoes what Constant (1980, 1984) pointed out for the automotive and turbojet traditions of practice in Germany and Britain. Community membership for both traditions was not mutually exclusive, because actors such as BMW and Rolls-Royce played an active role in both traditions of practice, brokering new technology and moving ideas and experiences from the old to the new technological communities and vice versa.

The notion of a technological community is therefore an ideal concept for making a positive contribution to innovation theories of socio-technical and techno-economic change, by demonstrating the value of multiple memberships and technological brokering (Hargadon 2003) between pre-existing and new technological traditions of practice, thus providing the missing link between notions of revolution and evolution of new technology and associated technological communities and traditions of practice. Technological revolution such as, for example, the cost-efficient manufacturing of the first IC by Fairchild Semiconductor, occurred as an evolutionary process of technological brokering in pre-existing traditions of practice. It was the result of multiple memberships of Fairchild, in particular of Jean Hoerni, Robert Noyce and Jay Last, in both the pre-existing technological communities related to transistors, and the new technological community of ICs. The same applies to the origins of GIS technology in Greece, where the innovators maintained one foot in the pre-existing surveying engineering and spatial planning traditions of practice, and the other in the new GIS community in Greece, shaping the evolution and trajectory of GIS technology in Greece according to certain application areas and issues informed by the relevant scientific theories and socio-cultural norms and systems of beliefs of the parent technological communities. Multiple memberships in technological communities, coupled with technological brokering, can therefore shed light on how revolutionary technologies come about from pre-existing technological communities and traditions of practice.

The concept of technological community, encompassing networks of individual adherents such as co-founders, co-directors and co-advisors, as well as teams of individuals, firms and other organisational actors, is an ideal theoretical construct to shift our thinking away from what Granovetter (2002: 36) in the new economic sociology calls the 'methodological individualism' of current neo-classical theories in mainstream economics. More importantly, technological communities and traditions of practice illustrate how micro-level socio-technical and techno-economic actions and decisions that people, teams and organisations make lead to macro-level structural changes at the industry level, which in turn affect future actions and decisions that actors make within the same industry and technological community, also building comparative advantages for some actors and their relevant social groups (Pinch and Bijker 1987). For example, the investigation of the origins and evolution of the SV semiconductor community shows how decisions that co-founders made, for example for patenting the first IC in 1959, or co-founding SV semiconductor firms, like Intel in the late 1960s, led to new socio-technical and techno-economic structures within the industry, which in turn have influenced the future developments of the technological tradition of the semiconductor community in SV and beyond. The same applies to the Greek GIS community where relevant social groups, such as the surveying engineering teams, have constructed a relative advantage for their tradition of practice compared to other relevant groups such as spatial planners, for absorbing resources and translating GIS innovations along a certain path of applications areas such as cadastre, tax and parcel management.

The study of the Greek GIS community can also be related to the debate regarding the loosening of the technology–society divide. As Rogers (1993: 21) has shown, diffusion should be seen as the reinvention of the technology by local actors who attach meaning to GIS innovations both at local and national scales, while Latour (2005) argues that a theory of socio-technical change should conceptualise the diffusion of innovations within a social system as a translation process in which different actors who make up the system deploy human and non-human elements in the formation of a complex socio-technical network which gradually re-defines a new technology. To make sense of this point using the analysis in Chapter 5, GIS teams that belong to a broad range of institutional settings, but share the dominant surveying engineering tradition of practice, have adopted, implemented and used a great deal of hardware, software and data in Greece. Equally in the 1990s and early 2000s they have also translated GIS according to their interests, shifting the emphasis of the debate to particular types of GIS applications (e.g. cadastre, parcel management, tax assessment) and specific GIS issues (e.g. geometry, accuracy in GIS databases, digital topographic data production) informed by the scientific knowledge, professional culture and beliefs associated with their pre-existing tradition of practice. In this sense they have enrolled members of other relevant social groups and technological communities related to geographic information handling and analysis in their agenda defining what GIS can (and cannot) be used for within a whole country.

Such a socio-technical theory of GIS diffusion takes into account a mixture of social, technological and economic purposes as well as a whole range of contextual arrangements at different scales of analysis. Thus it can be applied not only within different areas of the same country but also between different countries, facilitating the shift from a technological deterministic to a social constructivist view of GIS, or indeed any other technological innovation. The positioning of the actors in terms of structural equivalence (Burt 1987, 1992) as well as the patterns of direct GIS linkages within and across different relevant groups, can give insights into the structure of an emerging GIS community with respect to different social constructs such as institutional setting and disciplinary background (Assimakopoulos 2000). This SNA methodology can also be used for larger data sets within a much larger country than Greece, such as the United Kingdom or the United States, or even at an EU scale. If this procedure is repeated in more than one cycle it is also possible to study how the patterns of network linkages change over time and at different points in the diffusion process.

Similarly the semiconductor technology in SV can be seen as a technological community and tightly knit network encompassing more than a hundred semiconductor firms by the mid-1980s. The positioning of founders and co-founded semiconductor firms in SV according to structural equivalence or/and the patterns of direct cohesive ties among actors show how the semiconductor technology was embedded and shaped by a fast growing and tightly knit social network of people and firms. This network community of co-founded firms and co-founders as it was analysed and visualised in Chapter 6 has maintained close links with academia, as, for example, professors of electrical engineering and computer science at Stanford University have demonstrated clear leadership and corporate involvement with respect to co-founding and co-directing entrepreneurial firms in SV and beyond.

7.4 Implications for policy and practice

Innovation and technology policy has moved on in the last two decades based on the non-linear and fifth-generation innovation models (Rothwell 1992; Dodgson 2000) presented in Chapter 3. For example, the IST programme, which replaced Esprit in the fifth Framework programme (1998–2002) and has continued as a priority area in the current sixth one (2003–6), is very much market-driven and user-driven. Market pull has replaced technology push and the contrived notion of pre-competitive research, which did not survive to see the end of Esprit any more than did the dominance of hardware over software. And yet the EC's insistence on collaboration in much larger project networks and networks of excellence is as strong as ever in the ongoing sixth Framework programme. Collaboration in IST is still justified in the terms in which it has been justified in Esprit over the last two decades. Such justification is unlikely to be questioned in as much as collaboration among firms is hardly going out of fashion, though it commonly takes the growth form of joint ventures, mergers and acquisitions these days.

As the analysis in Chapter 4 suggests, mere inclusion in an Esprit formal collaboration did not guarantee that all participants actually took part in project networks, or that they contributed or benefited at all from the informal and personal networks uncovered in the 10 innovative case studies across a broad range of European ICT research. The reality of collaboration can be to a large extent the same old groupings and little new blood among the participants in Esprit and IST consortia regardless of their size. Though the EC has justified its requirement for collaboration among participants in its RTD programmes in terms of the advantages for innovation, collaboration has also satisfied the EC's own political requirements for involving SMEs and the European less favoured regions in its premier RTD programmes. Much Esprit collaboration was nominal in its early days in that it was arranged to satisfy application requirements, to improve prospects of funding, or to please project officers. In these circumstances, ICT innovation in Europe could hardly depend on the information readily available in the formal collaborations of Esprit.

This research indicates that much of the information for innovation in Esprit came from external sources – external to Esprit projects and often external to Europe. Very often it was acquired by personal and informal means via unanticipated partners in the most dynamic regions of the United States and worldwide. It would seem that the formality of collaboration in Esprit managed to accommodate this informal networking, not because the EC was sensitive to the importance of these networks and anxious not to disrupt their operation, but because their members found these useful and beneficial for ICT innovation and were determined that the EC would not interfere with their personal networks.

> The reasons the European Commission have to impose some partners is that they will be left out if they don't, and they put money into the pot in

> Europe, and occasionally they are saying why don't you pick up this company in trouble ... Yeah, all right we will have them in the project ... It is a pain but we did it because it helps ... The EC is full of politics. Full of it, and we try and avoid that, and try and focus rather hard on what we try to do.
>
> (ARM project manager)

The personal nature of most of these informal innovation networks highlights the need to recognise and accommodate individuals as much as formal legal entities such as firms and other knowledge-generating organisations. Personalisation of contractual and innovation activities is commonplace, for example in academia, where professors hold personal contracts with supporting government or industry agencies, as well as formal positions on board of directors or advisors, besides being co-founders in entrepreneurial ventures fuelled by university–industry interactions. The increasingly tacit and embedded nature of knowledge in personal networks and 'glocalised' communities should make personalisation one of the key issues for building a competitive strategy for firms and innovation policy for government agencies such as the EC and its RTD programmes (Assimakopoulos *et al.* 2000). Technological brokering and multiple memberships in technological communities and traditions of practice increasingly depend on key individuals and their teams and networks as much as formal organisations, as the research above has shown with regard to the Esprit project networks and the technological communities in Greece and SV.

Non-European firms may now participate in EC-supported programmes, but as non-funded and therefore unequal partners. This is some concession to reality, but still inadequate recognition of the non-European contribution to EC programmes in ICT RTD. The EC still requires European firms to collaborate so that they may be more efficient in IST research, more innovative, and thus more competitive, especially with respect to the Americans and Japanese (Assimakopoulos, Piekkari and Macdonald 2004). Such a notion is really no longer appropriate in the modern ICT industry, an industry whose products, structure, ownership, research, innovation and market are utterly global. It is positively surreal in a research programme like IST, which specifically seeks to exploit networks and clustering, and in the very ICT technology which facilitates information networking, both formal and personal. The consequence of the EC's continued insistence on European collaboration may well be reduced ICT RTD activity in Europe, and this is far too great a price to pay for the political convenience of the EC and its bureaucracy.

If there is one fundamental lesson to be highlighted for European strategy and policy makers from the continuing success of SV as the most innovative high-technology cluster in the world today, it is this privately led, personal and informal as much as community-like organisational model, that arose firstly in the semiconductor community in SV within Fairchild and its spin-offs, and has been reinforced with networked individualism in the past few decades, thus breeding continuous success for SV and related global technology regions as far apart as Israel and Taiwan, and generating over the past decade a 'glocalised' ecology of innovation and entrepreneurship, generation after generation of ICT RTD work. This informal community-like organisational model is also reflected in the personal networks and communities of customers and partners of some of the most innovative of the main UK Esprit contractors, in successful hardware and software firms such as ARM and APM in Cambridge, Flomerics in Surrey, Delcam in Birmingham, and other organisations such as ALCS in London. But the evidence from the Esprit case studies is that this reality has still to be concealed to some extent at the European level, as the EC still maintains a Euro-centric policy, trying to prevent American, Japanese and other non-European partners being considered as fully equal project partners for the current IST programme.

Despite such an inward-looking technology policy from the EC, it is worth speculating that as a result of the increasing globalisation of time-based RTD activity, spiralling costs and risks, technological innovation in ICT is increasingly going to be organised, by firm strategy and government policy makers alike, through transient network forms of organisation, clusters of projects, and associated technological communities across all sorts of boundaries – organisational, geographical, disciplinary, and so on. Such technological communities and personal networks will play a crucial role in giving meaning to networked or extended enterprises, taking stock of synergies and complementarities among research partners, reducing and spreading the associated RTD risks and costs among large and small firms, government agencies and universities and, last but not least, bringing together talented people and teams who create technological knowledge and opportunities across institutional and geographical boundaries.

From the outset, networking and collaboration in RTD work, as well as the social construction of new ICT technologies, will therefore be impossible to be studied without an understanding of personal network communities and new technological communities and traditions of practice that will increasingly shape and give meaning to diffusion and translation processes at a variety of spatial and institutional levels. Throughout the 2000s such networked communities will be 'glocalised', to some extent without propinquity, where collaborative links will simultaneously operate locally within CoPs, and across large geographical distances through NoPs, producing occupational and virtual communities (Assimakopoulos and Yan 2006b) bounded by common technological practices and interests. In the current and future RTD programmes supported by the EC the value of personal networks of innovators and personal 'glocalised' communities therefore ought to be recognised, institutionalised and nurtured down to the individual level, in order to foster European innovation and competitiveness in key RTD areas such as the IST priority area. This process of institutionalisation should become a key objective of both industry and governments in the European Union for fostering information societies and knowledge economies of the future, and for bringing Europe nearer to achieving its early 2000s Lisbon and Barcelona agendas, i.e. becoming the most competitive knowledge economy in the world by 2010.

7.5 Implications for further research

The starting point of this research was the need to investigate not only the role of personal networks in ICT innovation, but also the origins and early critical stages of development for technological communities and traditions of practice with regard to GIS innovations in Greece and semiconductors in SV. It was also felt important to study the dynamics of the SV semiconductor community from the 1960s onwards, and the diffusion and adoption of GIS innovations in Greece from the early 1980s onwards. Further research is required in terms of the dynamics of the SV semiconductor community for the past two decades, and more specifically after 1986 when the SEMI data set stops providing data about founders and co-founded semiconductor firms in SV. The same applies for further research in the evolution of the Greek GIS community after the mid-1990s, for at least two related reasons.

First, it is expected that many more GIS teams from an increasing number of geographical locations and institutional and disciplinary settings have adopted and implemented GIS software, playing an active part in the development of this GIS community in the late 1990s and in the early 2000s. These actors are likely to highlight even more clearly the essential complementarities of interests between the different components of this new technological community. They are also likely to support the argument that the cumulative number of GIS adopters follows the S-shaped logistic curve as proposed by Rogers (2003) in his theoretical model of the diffusion of innovations and by Crane (1972) in her theoretical model of the growth of invisible colleges in science.

Second, the complementarities and inter-dependencies between the various stakeholders were explored above with respect to the patterns of linkages within and across the various relevant social groups who make up the Greek GIS community in a generalised way. To demonstrate the value of computerised SNA in understanding the heterogeneous structure of an emerging technological community, future research needs to consider a number of issues related to a more detailed analysis of linkages. Instead of identifying components or relevant social groups of a GIS community, it makes sense to go back to such clusters and ask each GIS team to provide information about its manifold linkages with the rest. This could enrich the analysis at least in three respects.

First, it could untangle the multiple strands of GIS linkages. Economic or contractual relations are often different in nature from knowledge and social relations of friendship or/and advice. The more multiplex a GIS linkage between two GIS teams the more durable is likely to be the interdependency and complementarities with respect to GIS innovations. Second, it could shed light in the development of power relations. Many relations of a formal nature work two ways but reciprocity is not always the case in the processes of innovation adoption and implementation. Asymmetric relations based on one-way sharing or exchange of information and other scarce resources, such as knowledge, characterise many informal and personal linkages. Third, it could show the strength of each linkage with respect to various measures such as centrality of actors, accessibility or frequency of ties.

Instead of a single, binary and symmetric co-founder semiconductor or GIS socio-matrix, future research into the emergence of technological communities could analyse multiple, directional and valued socio-matrices which can yield a much more detailed picture of the heterogeneous nature of a technological community throughout a whole region or country. In particular socio-metric techniques might be used to estimate the social distances between different actors and relevant groups who give birth to a new technological tradition of practice or adopt and implement new software. This analysis could be particularly interesting in demonstrating how different people, teams and groups are positioned relatively to each other in a two (or more) dimensional space of technological specialisation. It could also provide insights into how the patterns of linkages change at different points in the diffusion process of new software or organisational practices related to specific innovations.

With these considerations in mind a few particularly interesting questions in relation to the further development of the SV semiconductor and Greek GIS communities need to be studied:

- Will the significance of Fairchild and 'Fairchildren' spin-offs such Intel and Intersil maintain their centrality in the structure of the semiconductor community in SV in the late 2000s?
- Will the patterns of connections within and between government, academia and private sector teams remain similar with those discussed in relation to the Greek GIS community in the late 2000s?
- Will the GIS teams that share a surveying engineering background maintain their primacy in the inner circle in the Greek GIS community?

A potentially fruitful area of research arising out of these investigations is also how network structure and culture affect each other. According to Erickson (1996), culture shapes the development of social networks underlying network phenomena, such as the diffusion of GIS innovations, as well as the formation of cohesive subgroups in social systems such as the SV semiconductor community. Figure 7.1 shows a socio-technical pyramid which links different elements of culture, structure and socio-technical networks together. This book has adopted a bottom-up approach, while more

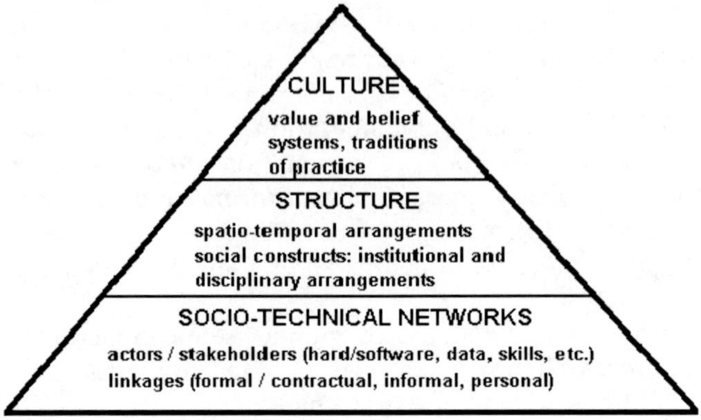

Figure 7.1 A socio-technical pyramid for investigating the concept of a technological community and related tradition of practice.

sophisticated models of GIS and other technological communities and emerging traditions of practice might explicitly address all three levels of analysis from a top-down perspective.

For example they might investigate how systems of values and beliefs within existing scientific and technological communities affect the emergence of different new GIS technological traditions of practice within different countries. An interesting research avenue could therefore be to investigate the effect that different geographic information handling cultures have in the emergence of different GIS communities in various countries. As noted above, in Greece and Germany the surveying engineering tradition of practice influences to a large extent the institutional and disciplinary context within which GIS technology is embedded at a national scale. In contrast, the English-speaking world (i.e. the United Kingdom, the United States) is more influenced by the physical and human geography traditions, while other countries such as the Netherlands have one foot in the surveying engineering camp and the other in the geography camp.

The findings of this research also indicate the vital role that European Union funding has played in facilitating the diffusion of technological innovations, such as GIS software, in the less favoured regions of the European Union such as Greece. This demands further research on the impacts of such funding on various groups and actors at the EU scale in conjunction with the findings on the informal personal networks of the main UK Esprit contractors, who would prefer to collaborate with their US and worldwide unofficial partners rather than their formal partners in the less favoured EU regions in order merely to fulfil the criteria of innovation policies supported by the EC. A fruitful research avenue may therefore be to explore the links between EC funding to the stage of development of such RTD networks and communities. For example, the findings of the analysis in Chapter 5 suggest that EC funding is of particular importance in the early critical stages of GIS diffusion. However, more detailed investigation is required to ascertain the need for similar types of funding in subsequent stages and to identify the point at which the diffusion process becomes self-sustaining as a result of the critical mass formation.

Finally, the evolution and dynamics of technological communities could be investigated using recent network topology models, such as 'small world' and 'scale free' networks introduced by physicists and mathematicians (Watts and Strogatz 1998; Barabasi and Albert 1999; Newman 2001) which are generating considerable interest from social scientists and innovation scholars alike (Granovetter 2003; Fleming, King and Juda 2004; Schilling and Phelps 2004; Yan and Assimakopoulos 2006). SNA modelling of the network topology that underlies emerging technological communities and personal networks, with respect to such key issues as the role of 'small world' and 'scale free' networks in distributed RTD work, seems a fruitful avenue for research in time-based innovation of the future.

Appendix

SNA concepts and metrics

The computerised network analysis above made use of three software packages for social network analysis and visualisation: Ucinet 6 (Borgatti, Everett and Freeman 2002), Mage 5.2 (Richardson and Presley 2001) and Pajek (Batagelj and Mrvar 2004). Ucinet was used in Chapter 4 to compute a set of coordinates for the personal network of each Esprit technological leader following a three-step approach. It placed all nominations within a binary socio-matrix (1 indicates the existence of a connection, 0 indicates the lack of a connection), revealing who was connected with whom within a particular Esprit project. An assumption was made that all ties were reciprocal in nature since nearly all respondents indicated that they supplied information for innovation to others of more or less equal value. Second, it calculated Euclidian distances among the nominated individuals (Burt 1987; Wasserman and Faust 1994). Euclidian distance is a measure of structural similarity among the nodes of a network as is explained in greater detail below. Third, based on Euclidian distances, a set of (x, y, z) coordinates for each individual was calculated using a three-dimensional scaling routine (Borgatti, Everett and Freeman 2002). Based on each set of coordinates, Mage produced three-dimensional kinetic images for exploring the social structure of each personal network. It is pertinent that Mage was initially produced for the visualisation of protein molecules, but has since been used to visualise and make sense of social structures (Freeman, Webster and Kirke 1998).

It is worth noting here that one of the early advocates for using structural equivalence as a measure of innovative behaviour is Burt (1987) who argued that social contagion and thus innovation is more likely to occur when actors have similar patterns of connections within a network (structural equivalence) than when they have direct contact (cohesion). According to Wasserman and Faust (1994: 356) two actors are structurally equivalent if they have mathematically identical connections to and from all other actors in a network. In practice it is unlikely that any actors in a social system will be exactly equivalent. As a result a measure of structural equivalence is needed so that the extent to which actors are equivalent can be shown. In the following equation structural equivalence is measured by Euclidian distance (Burt 1987).

$$d_{ij} = \sqrt{\sum_{k=1}^{g}[(x_{ij} - x_{jk})^2 + (x_{ki} - x_{kj})^2]} \qquad \text{for } i \neq k, j \neq k \tag{1}$$

The calculation of Euclidian distances follows Equation 1 and is based on the socio-metric representation of a network as a square socio-matrix consisting of g actors. As a result k counts from actor 1 to actor g, while x_{ik} represents the value of the tie from actor i to actor k. If actors i and j are structurally equivalent, then the entries in their respective rows and columns of the socio-matrix will be identical and thus the Euclidian distance (d_{ij}) between them will be equal to zero. To the extent that two actors are not structurally equivalent, the Euclidian distance between them will be large. Euclidian distance has the properties of a distance socio-matrix. Such a socio-matrix can be computed for Equation 1 using specialised software such as Ucinet 6 (Borgatti, Everett and Freeman 2002).

In contrast to the notion of structural equivalence which takes into account the whole pattern of connections in a network, cohesion is based only on the direct connections between actors. According to Wasserman and Faust (1994: 251), the literature on cohesive subgroups has used various criteria to conceptualise the idea of a subgroup in a network. Three types of criteria are usually used to define cohesive subgroups. These are: mutuality of ties, closeness or accessibility of subgroup members, and frequency of ties among subgroup members compared to non-members. A cohesive subgroup, i.e. clique, is often defined as a fully connected subgroup of a minimum of three actors.

The centrality of actors in a social system was also one of the first concepts studied by network scholars for explaining innovative behaviour or how information of value to innovation flows in a network (Freeman 1979). In the diffusion of innovations field it is usually claimed that individuals or other adopting units who maintain many ties with others in a social system are more likely to be early adopters of new ideas and play the role of opinion leader for the rest in the system (Rogers 1983: 277). Although it is rather a crude measure of real prominence, the number of nominations received by the other members of a network, or the in-degree centrality, is the most commonly used measure of opinion leadership in a network (Freeman 1979). In-degree centrality is also called degree prestige (Wasserman and Faust 1994: 202). On the other hand, out-degree centrality is defined as the number of nominations sent by each member to all others in the network. From a conceptual viewpoint the concern with centrality also stems from two quite different structural situations. As Freeman, Borgatti and White (1991: 141) note, actors are central in a network if they are close to the other members of the network, or because they stand between others on the path of communication. As a result of these structural considerations it is also necessary to consider closeness centrality, betweenness and flow betweenness centrality.

Closeness centrality is the extent to which an actor is near other actors in a network (Valente 1995: 52). An actor with high closeness centrality reaches others in the network more quickly, through fewer intermediaries, than an actor with a lower centrality score. Closeness centrality measures how many steps, on average, it takes for an actor to reach everyone else in the network. According to Wasserman and Faust (1994: 184), closeness is equated with minimum distance. As an actor grows farther apart from other actors, their centrality will decrease. Conversely, actors who are close to others can be very productive in communicating information to these other actors and may take on the role of opinion leaders.

Betweenness centrality is a measure of how often an actor lies in the geodesic or shortest path linking two other actors of a network (Valente 1995: 52). Freeman, Borgatti and White (1991: 151), however, proposed the term 'flow betweenness centrality' because they point out that there is no reason to believe that actors in a social system restrict their communication to the shortest paths in their networks. Centrality flow betweenness considers all independent paths not only the shortest paths. High centrality betweenness or flow betweenness indicates that the actor is an intermediary between many others in the network, a possible link between many potential nodes in the network, and that s/he may broker many relationships. Actors who are high in either of these centrality measures (closeness or betweenness/flow betweenness) are more likely to receive information concerning an innovation relatively earlier than others in the network and thus may adopt new ideas influencing (positively or negatively) the others in the system earlier (Burt 2004). Therefore both of these centrality measures are usually associated with structural holes (Burt 1992) and opinion leadership in an informal innovation network.

Glossary of SNA terminology

(The definitions are based on Wasserman and Faust 1994.)

Betweenness centrality A measurement of centrality, indicating how 'powerful' an actor is in terms of controlling information flow in a network. The idea here is that an actor is central if s/he lies between other actors on the shortest paths connecting these actors.

Centralisation An index at group level, measuring how variable or heterogeneous the actor centralities are. It records the extent to which a single actor has high centrality and the other actors in the network have low centrality.

Centrality An index used to indicate how critical an actor is in a network. Degree is the most popular way of measuring centrality. *See also* **betweenness centrality** and **closeness centrality**.

Clique A maximal complete subgroup in which all actors are directly connected to each other, and there are no other actors that are directly connected to all members of the clique.

Closeness centrality A measurement of centrality indicating how close an actor is to all other actors in a network. The idea here is that an actor is central if s/he can reach many other actors in a network in a few steps.

Degree An index measured by the number of linkages adjacent to an actor.

Density An index used to indicate how actors are closely or loosely connected in a network. It is measured by the proportion of possible linkages that are actually present in a network.

Euclidean distance An index to measure structural similarity between actors of a network. The less two actors are structural equivalent, the larger the Euclidean distance between them. *See also* the Appendix.

Multi-dimensional scaling A way of visualising Euclidean distances. Networks are often visualised by two- or three-dimensional scaling in a graphical way with (x, y, z) coordinates, presenting a map of geometrical 'Euclidian' distances between actors in a network.

Position A group of actors who have similar pattern of relations to all other actors in a network. People in a similar social position are more likely to have similar profiles, i.e. social activities, ties and interactions, than people in different positions in a network.

Structural equivalence Two actors are structurally equivalent if they have mathematically identical connections (structural similarity) to and from all other actors in a network (*see also* the Appendix). The actors who occupy the same social positions are said to be structurally equivalent.

References

Aldrich, H. and von Glinow, M., 1992, Personal Networks and Infrastructure Development, in D. Gibson *et al.* (eds), *Technopolis Phenomenon*, New York: Rowman and Littlefield, 125–45.

Allen, T. J., 1977, *Managing the Flow of Technology*, Cambridge, MA: MIT Press.

Arora, A. and Gambardella, A., 1990, Complementarity and External Linkages, *Journal of Industrial Economics*, 38, 361–79.

Assimakopoulos, D., 1996, Greece: The Development of a GIS Community, in I. Masser, H. Campbell and M. Craglia (eds), *GIS Diffusion*, London: Taylor and Francis, 147–61.

——, 1997a, The Greek GIS Community, PhD thesis, Department of Town and Regional Planning, University of Sheffield.

——, 1997b, GIS Diffusion in Greece, in M. Craglia and H. Couclelis (eds), *Geographic Information Research: Bridging the Atlantic*, London: Taylor and Francis, 111–28.

——, 1997c, Social Networks and GIS Diffusion in Greece, in C. Capineri and P. Rietveld (eds), *Networks in Transport and Communications: A Policy Approach*, Aldershot: Ashgate, 153–74.

——, 2000, Social Network Analysis as a Tool for Understanding the Diffusion of GIS Innovations, *Environment and Planning B*, 27, 627–40.

—— (ed.), 2003, Introduction, double special issue on Collaboration and Networking, *International Journal of Technology Management*, 25, 1–4.

Assimakopoulos, D. and Macdonald, S., 1999, Collaboration and Innovation Networks in Esprit, *Prometheus*, 17, 299–308.

——, 2002, A Dual Approach to Understanding Information Networks, *International Journal of Networking and Virtual Organisations*, 1, 1–16 (reprinted in the *International Journal of Technology Management*, 25, 96–112).

——, 2003, Personal Networks and IT Innovation in the Esprit Program, *Innovation: Management, Policy and Practice*, 5, 15–28.

Assimakopoulos, D., and Kenney, M., 2005, Exploring the Relational Nature of Leadership in Professorial Entrepreneurship, paper presented at the 4th International Critical Management Studies Conference, held at the Judge Institute of Management, University of Cambridge.

Assimakopoulos, D. and Yan, J., 2005, Social Network Analysis and Communities of Practice, in E. Coakes and S. Clarke (eds), *Encyclopedia of Communities of Practice in Information and Knowledge Management*, London: Idea Group, 474–80.

——, 2006a, Sources of Knowledge Acquisition for Chinese Software Engineers, *R&D Management*, 36, 97–106.

——, 2006b, The External Linkages of Community of Practice: Integrating Communities Using Internet Technology Forums, in H. Sherif and T. Khalil, *Management of Technology: Selected Papers from the Thirteenth International Conference on Management of Technology*. Oxford: Elsevier, 199–221.

Assimakopoulos, D., Chrissafis, A., Gustavsson, P., Macdonald, S. and Marschan-Piekkari, R., 2000, *Exploiting Informal Information Flow in the IST Programme*, Final InEsprit (project 29538), Report, Brussels.

Assimakopoulos, D., Everton, S. and Tsutsui, K., 2003, The Semiconductor Community in Silicon Valley, *International Journal of Technology Management*, 25, 181–99.

Assimakopoulos, D., Piekkari, R. and Macdonald, S., 2004, Esprit: Europe's Response to US and Japanese Dominance in IT, in R. Coopey (ed.), *Information Technology Policy*, Oxford: Oxford University Press, 247–63.

Azad, B., 1997, Leading and Managing the GIS Implementation Process: Beyond the Myth of the GIS Champion, in M. Craglia and H. Couclelis (eds), *Geographic Information Research: Bridging the Atlantic*, London: Taylor and Francis, 146–64.

Badaracco, J., 1991, *The Knowledge Link*, Boston, MA: Harvard Business School Press.

Barabasi, A. L. and Albert, R., 1999, Emergence of Scaling in Random Networks, *Science*, 286, 509–12.

Barley, S. R., 1990, The Alignment of Technology and Structure through Roles and Networks, *Administrative Science Quarterly*, 35, 61–103.

Barney, J. B., 1991, Firm Resources and Sustained Competitive Advantage, *Journal of Management Studies*, 17, 99–120.

Batagelj, V. and Mrvar, A., 2004, *Pajek: Program for Large Network Analysis*, available at: http://vlado.fmf.uni-lj.si/pub/networks/pajek (accessed 10 June 2004).

Bell, C. and Newby, H., 1971, *Community Studies: An Introduction to the Sociology of the Local Community*, London: Allen and Unwin.

Berlin, L., 2005, *The Man Behind the Microchip: Robert Noyce and the Invention of Silicon Valley*, Oxford: Oxford University Press.

Bernard, H. R., Killworth, P., Kronenfeld, D. and Sailer, L., 1984, The Problem of Informant Accuracy: The Validity of Retrospective Data, *Annual Review of Anthropology*, 13, 495–517.

Bijker, W. E., 1995, *Of Bicycles, Bakelites and Bulbs: Toward a Theory of Socio-technical Change*, Boston, MA: MIT Press.

Bijker, W. E. and Law, J. (eds), 1992, *Shaping Technology / Building Society*, Boston, MA: MIT Press.

Bijker, W. E, Hughes, T. P. and Pinch, T. F. (eds), 1987, *The Social Construction of Technological Systems: New Directions in the Sociology and History of Technology*, Boston, MA: MIT Press.

Blau, P. M., 1974, Parameters of Social Structure, *American Sociological Review*, 39, 615–34.

Blosch, M. and Preece, D., 2000, Framing Work through a Socio-technical Ensemble: The Case of Butler Co., *Technology Analysis and Strategic Management*, 12, 91–102.

Boisot, M., 1998, *Knowledge Assets*, Oxford: Oxford University Press.

Bonacich, P., 1987, Power and Centrality: A Family of Measures, *American Journal of Sociology*, 92, 1170–82.

Borgatti, S.P., Everett, M.G. and Freeman, L.C., 2002, *Ucinet 6.0 for Windows: Software for Social Network Analysis*, Cambridge, MA: Analytic Technologies.

Braun, E. and Macdonald, S., 1982, *Revolution in Miniature: The History and Impact of Semiconductor Electronics*, Cambridge: Cambridge University Press.

Bresnahan, T. and Gambardella, A. (eds), 2004, *Building High-tech Clusters: Silicon Valley and Beyond*, Cambridge: Cambridge University Press.

Brown, J. S. and Duguid, P., 1991, Organizational Learning and Communities of Practice, *Organization Science*, 2, 40–57.

——, 1998, Organizing Knowledge, *California Management Review*, 40, 90–112.

——, 2000a, *The Social Life of Information*, Boston, MA: Harvard Business School Press.

——, 2000b, Mysteries of the Region: Knowledge Dynamics in Silicon Valley, in C. M. Lee, W. Miller, M. Gong-Hancock and H. Rowen (eds), *The Silicon Valley Edge*, Stanford, CA: Stanford University Press, 16–39.

——, 2001, Knowledge and Organization: a Social-Practice Perspective, *Organization Science*, 12, 198–213.

Browning, L. D. and Shetler, J. C., 2000, *Sematech: Saving the U.S. Semiconductor Industry*, College Station, TX: Texas A&M University Press.

Bryman, A., 1988, *Quantity and Quality in Social Research*, London: Unwin Hyman.

——, 2004, *Social Research Methods*, Oxford: Oxford University Press.

Burt, R. S., 1987, Social Contagion and Innovation: Cohesion versus Structural Equivalence, *American Journal of Sociology*, 92, 1287–1335.

——, 1992, *Structural Holes: The Social Structure of Competition*, Boston, MA: Harvard University Press.

——, 2004, Structural Holes and Good Ideas, *American Journal of Sociology*, 110, 349–99.

——, 2005, *Brokerage and Closure*, Oxford: Oxford University Press.

Cabinet Office, 1999, e-commerce@its.best.uk (accessed 10 June 2000).

Callon, M., 1993, Variety and Irreversibility in Networks of Technique Conception and Adoption, in D. Foray and C. Freeman (eds), *Technology and the Wealth of Nations*, London: Pinter, 232–68.

Callon, M. and Latour, B., 1981, Unscrewing the Big Leviathan, in K. Knorr-Cetina and A. V. Cicourel (eds), *Advances in Social Theory and Methodology: Toward an Integration of Micro and Macro Sociologies*, London: Routledge, 277–303.

Callon, M. and Law, J., 1982, On Interests and their Transformation: Enrolment and Counter-Enrolment, *Social Studies of Science*, 12, 615–25.

Campbell, H. and Masser, I., 1995, *GIS and Organisations: How Effective are GIS in Practice?*, London: Taylor and Francis.

Carayannis, E. and Alexander, J., 1999, Winning by Co-operating in Strategic Government – University – Industry Partnerships: The Power of Complex, Dynamic Knowledge Networks, *Journal of Technology Transfer*, 24, 197–210.

Castells, M., 1996, *The Rise of the Network Society*, Oxford: Blackwell.

——, 2001, *The Internet Galaxy*, Oxford: Oxford University Press.

Castilla, E. J., 2003, Networks of Venture Capital Firms in Silicon Valley, *International Journal of Technology Management*, 25, 113–35.

Castilla, E. J., Hwang, H., Granovetter, H. and Granovetter, M., 2000, Social

Networks in Silicon Valley, in C. M. Lee, W. F. Miller, M. G. Hancock and H. S. Rowen (eds), *The Silicon Valley Edge: A Habitat for Innovation and Entrepreneurship*, Stanford, CA: Stanford University Press, 218–47.

Charles, D. R. and Howells, J., 1992, *Technology Transfer in Europe: Public and Private Networks*, London: Belhaven Press.

Clark, B., 1998, *Creating Entrepreneurial Universities: Organizational Pathways of Transformation*, New York: Pergamon Press.

Coleman, J. S., 1988, Social Capital in the Creation of Human Capital, *American Journal of Sociology*, 94, 95–120.

Coles, A. M., Harris, L. and Dickson, K., 2003, Testing Goodwill: Conflict and Cooperation in New Product Development Networks, *International Journal of Technology Management*, 25, 51–64.

Constant, E. W., 1980, *The Origins of Turbojet Revolution*, Baltimore, MD: Johns Hopkins University Press.

——, 1984, Communities and Hierarchies: Structure in the Practice of Science and Technology, in R. Laudan (ed.), *The Nature of Technological Knowledge: Are Models of Scientific Change Relevant?*, Dordrecht: Reidel, 27–46.

——, 1987, The Social Locus of Technological Practice: Community, System, or Organisation?, in W. E. Bijker, T. P. Hughes and T. J. Pinch (eds), *The Social Construction of Technological Systems*, Cambridge, MA: MIT Press, 223–42.

Conway, S., 1995, Informal Boundary-spanning Communication in the Innovation Process: An Empirical Study, *Technology Analysis and Strategic Management*, 7, 327–42.

——, 1997, Strategic Personal Links in Successful Innovation: Link-pins, Bridges, and Liaisons, *Creativity and Innovation Management*, 6, 226–33.

Conway, S. and Steward, F., 1998, Mapping Innovation Networks, *International Journal of Innovation Management*, 2, 223–54.

Coombs, R. and Metcalfe, S., 1998, Distributed Capabilities and the Governance of the Firm, CRIC Discussion Paper 16, University of Manchester, available at: http://les1.man.ac.uk/usercgi/cric/cricpaperdl.asp?paper=dp16 (accessed 14 January 2004).

Coombs, R., Richards, A., Saviotti, P. P. and Walsh, V. (eds), 1996, *Technological Collaboration: The Dynamics of Cooperation in Industrial Innovation*, Cheltenham: Edward Elgar.

Coombs, R., Harvey, M. and Tether, B., 2001, Analysing Distributed Innovation Processes, CRIC Discussion Paper 43, University of Manchester, available at: http://les1.man.ac.uk/usercgi/cric/cricpaperdl.asp?paper=dp43 (accessed 14 January 2004).

Coppock, T. and Rhind, D. W., 1991, The History of GIS, in D. J. Maguire, D. W. Rhind and M. F. Goodchild (eds), *Geographical Information Systems: Principles and Applications*, London: Longman, 21–43.

Couclelis, H., 2004, The Third Domain: The Spread and Use of GIS within Social Science, *Cartographica*, 39, 17–24.

Crane, D., 1972, *Invisible Colleges: Diffusion of Scientific Knowledge*, Chicago: University of Chicago Press.

Croon, L., 1939, *Das Fahrrad und seine Entwicklunk*, Berlin: VDI-Verlag.

Cross, R. and Prusak, L. 2002, The People Who Make Organizations Go – or Stop, *Harvard Business Review*, 80, 104–12.

Cross, R., Prusak, L. and Parker, A., 2001, Knowing what we Know: Supporting

Knowledge Creation and Sharing in Social Networks, *Organizational Dynamics*, 30, 100–120.

Cross, R., Borgatti, S. and Parker, A., 2002, Making Invisible Work Visible: Using Social Network Analysis to Support Strategic Collaboration, *California Management Review*, 44, 25–46.

Davenport, T. H. and Prusak, L., 1998, *Working Knowledge: How Organizations Manage What They Know*, Boston, MA: Harvard Business School Press.

DeCarolis, D. M. and Deeds, D. L., 1999, The Impact of Stocks and Flows of Organizational Knowledge on Firm Performance: An Empirical Investigation of the Biotechnology Industry, *Strategic Management Journal*, 20, 953–69.

De Nooy, W., Mrvar, A. and Batagelj, V., 2005, *Exploratory Social Network Analysis with Pajek*, Cambridge: Cambridge University Press.

Dickson, D., 1983, The New Push for European Science Cooperation, *Science*, 10 June, 1134–36.

Dodgson, M., 1993, Learning, Trust and Technological Collaboration, *Human Relations*, 46, 77–95.

——, 2000, *The Management of Technological Innovation: An International and Strategic Approach*, Oxford: Oxford University Press.

Doreian, P., 1995, Social Network Analysis as a Scientific Revolution: Thinking in Circles or Genuine Progress?, Keynote Address at the 15th Sunbelt and 4th European Network Conference, London, July.

Dunlop, J. B., 1888, An Improvement in Tyres of Wheels for Bicycles, Tricycles or Other Road Cars, British Patent 10607, Filed on July 23.

Dyer, J. H. and Singh, H., 1998, The Relational View: Cooperative Strategy and Sources of Interorganizational Competitive Advantage, *Academy of Management Review*, 23, 660–79.

e-Europe, 2002, available at: http://europa.eu.int/information_society/eeurope/2002/index_en.htm (accessed 15 May 2004).

Eisenhardt, K. M. and Martin, J. A., 2000, Dynamic Capabilities: What Are They? *Strategic Management Journal*, 21, 1105–21.

Erickson, B., 1996, The Structure of Ignorance, *Connections*, 19, 28–38.

Etzkowitz, H., 2003, Research Groups as 'Quasi Firms': The Invention of the Entrepreneurial University, *Research Policy*, 32, 109–21.

Etzkowitz, H. and Leydesdorff, L., 2000, The Dynamics of Innovation: From National Systems and 'Mode 2' to a Triple Helix of University – Industry – Government Relations, *Research Policy*, 29, 109–23.

Etzkowitz, H., Wester, A., Gebhardt, C. and Terra, B., 2000, The Future of the University and the University of the Future: Evolution of the Ivory Tower to Entrepreneurial Paradigm, *Research Policy*, 29, 313–30.

Everton, S. F., 2004, A Guide for the Visually Perplexed: Visually Representing Social Networks, Department of Sociology, Stanford University, available at: www.stanford.edu/group/esrg/siliconvalley/papers.htm (accessed 15 January 2004).

Ferrary, M., 2003, The Gift Exchange in the Social Networks of Silicon Valley, *California Management Review*, 45, 120–38.

Fleming, L., King, C. and Juda, A., 2004, Small Worlds and Innovation, unpublished paper, Harvard University Business School, available at: http://pine.hbs.edu/external/facPersonalShow.do?pid=6605 (accessed 15 December 2005).

Foss, N. J., 2002, New Organizational Forms: Critical Perspectives, *International Journal of the Economics of Business*, 9, 1–8.

Fox, S., 2000, Communities of Practice, Foucault and Actor-Network Theory, *Journal of Management Studies*, 37, 853–67.

Freeman, C., 1991, Networks of Innovators: A Synthesis of Research Issues, *Research Policy*, 20, 499–514.

Freeman, L. C., 1979, Centrality in Social Networks: Conceptual Clarification, *Social Networks*, 1, 215–39.

Freeman, L. C., Borgatti, S. P. and White, D. R., 1991, Centrality in Valued Graphs: A Measure of Betweenness based on Network Flow, *Social Networks*, 13, 141–54.

Freeman, L. C., Webster, C. M. and Kirke, D. M., 1998, Exploring Social Structure Using Dynamic Three-Dimensional Color Images, *Social Networks*, 20, 109–18.

Galinski, C., 1984, Information – the Basis of Japan's Forecast Technological and Economic Development, *ASLIB Proceedings*, 36, 24–60.

Galunic, D. C. and Eisenhardt, K. M., 2001, Architectural Innovation and Modular Corporate Forms, *Academy of Management Journal*, 44, 1229–49.

Gargiulo, M., 1993, Two Step Leverage: Managing Constraint in Organisational Politics, *Administrative Science Quarterly*, 39, 1–19.

Geisler, E., 1993, On the importance of University – Industry – Government Cooperation: A Global Perspective, *International Journal of Technology Management*, 8, 435–38.

Georghiou, L., 1992, *The Impact of European Community Policies for Research and Technological Development upon Science and Technology in the United Kingdom*, report prepared for DGXII, London, UKIMPACT.

——, 1999, Socio-economic Effects of Collaborative R&D: European Experiences, *Journal of Technology Transfer*, 24, 69–79.

——, 2001, Evolving Frameworks for European Collaboration in Research and Technology, *Research Policy*, 30, 891–903.

Gherardi, S., 2001, From Organizational Learning to Practice-based Knowing, *Human Relations*, 54, 131–39.

Gibbons, J. F., 2000, The Role of Stanford University, in C. M. Lee, W. F. Miller, M. G. Hancock and H. S. Rowen (eds), *The Silicon Valley Edge: A Habitat for Innovation and Entrepreneurship*, Stanford, CA: Stanford University Press, 200–217.

Gibbons, M., Limoges, C., Nowotny, H., Schwartzman, S., Scott, P. and Trow, M., 1994, *The New Production of Knowledge: The Dynamics of Science and Research in Modern Societies*, London: Sage.

Gillmor, S. C., 2004, *Fred Terman at Stanford: Building a Discipline, a University and Silicon Valley*, Stanford, CA: Stanford University Press.

Giusti, W. and Georghiou, L., 1988, The Use of Co-nomination Analysis in Real-time Evaluation of an R&D Programme, *Scientometrics*, 14, 265–81.

Goodchild, M. F., 1992, Geographical Information Science, *International Journal of Geographical Information Systems*, 6, 31–45.

Granovetter, M. S., 1973, The Strength of Weak Ties, *American Journal of Sociology*, 78, 1360–80.

——, 1982, The Strength of Weak Ties: A Network Theory Revisited, in P. V. Marsden and N. Lin (eds), *Social Structure and Network Analysis*, Beverly Hills, CA: Sage, 105–30.

——, 1985, Economic Action and Social Structure: The Problem of Embeddedness, *American Journal of Sociology*, 91, 481–510.

——, 2002, A Theoretical Agenda for Economic Sociology, in M. F. Guillen, R. Collins, P. England and M. Meyer (eds), *The New Economic Sociology: Developments in an Emerging Field*, New York: Russell Sage, 35–60.

——, 2003, Ignorance, Knowledge and Outcomes in a Small World, *Science*, 301, 773–74.

——, 2005, The Impact of Social Structure on Economic Outcomes, *Journal of Economic Perspectives*, 19, 33–50.

Gulati, R., 1995, Social Structure and Alliance Formation Patterns: A Longitudinal Analysis, *Administrative Science Quarterly*, 40, 619–52.

Gulati, R. and Gargiulo, M., 1999, Where Do Inter-organisational Networks Come From? *American Journal of Sociology*, 104, 1439–93.

Gutting, G., 1984, Paradigms, Revolutions and Technology, in R. Laudan (ed.), *The Nature of Technological Knowledge: Are Models of Scientific Change Relevant?*, Dordrecht: Reidel, 47–65.

Hagedoorn, J. and Schackenraad, J., 1993, A Comparison of Private and Subsidized R&D Partnerships in the European Information Technology Industry, *Journal of Common Market Studies*, 31, 373–90.

Hagedoorn, J., Link, A. N. and Vonortas, N. S., 2000, Research Partnerships, *Research Policy*, 29, 567–86.

Hamel, G., Doz, Y. L. and Prahalad, C. K., 1989, Collaborate with Your Competitors and Win, *Harvard Business Review*, 67, 133–39.

Hammersley, M. and Atkinson, P., 1983, *Ethnography: Principles in Practice*, London: Tavistock.

Hargadon, A., 2003, *How Breakthoroughs Happen: The Surprising Truth about How Companies Innovate*, Boston, MA: Harvard Business School Press.

Harvey, F., 2001, Constructing GIS: Actor Networks of Collaboration, *URISA Journal*, 13, 30–37.

Henton, D., 2000, A Profile of the Valley's Evolving Structure, in C. M. Lee, W. F. Miller, M. G. Hancock and H. S. Rowen (eds), *The Silicon Valley Edge: A Habitat for Innovation and Entrepreneurship*, Stanford, CA: Stanford University Press, 46–58.

Hildreth, P. and Kimble, C., 2000, Communities of Practice in the Distributed International Environment, *Journal of Knowledge Management*, 4, 27–38.

Hindle, T., 2003, The Third Age of Globalisation, in *The Economist: The World in 2004*, London: The Economist, 107–8.

Hoefler, D. C., 1971, Silicon Valley USA, *Electronic News*, 11 January, 1, 4–5.

House of Commons Trade and Industry Committee, 1988, *Information Technology*, London: HMSO.

Howells, J., 1999, Research and Technology Outsourcing and Innovation Systems: An Exploratory Analysis, *Industry and Innovation*, 6, 113–29.

Howells, J. and Nedeva, M., 2003, The International Dimension to Industry – Academic Links, *International Journal of Technology Management*, 25, 5–17.

Hughes, T. P., 1979, Emerging Themes in the History of Technology, *Technology and Culture*, 20, 697–711.

Innes, J. E. and Simpson, D. M., 1993, Implementing GIS for Planning: Lessons from the History of Technological Innovation, *Journal of American Planning Association*, 59, 230–36.

Kenney, M. (ed.), 2000, *Understanding Silicon Valley: The Anatomy of an Entrepreneurial Region*, Stanford, CA: Stanford University Press.

Kenney, M. and Goe, R. W., 2004, The Role of Social Embeddedness in Professorial Entrepreneurship, *Research Policy*, 33, 691–707.

Kimble, C., Hildreth, P. and Wright, P., 2001, Communities of Practice: Going Virtual, in *Knowledge Management and Business Model Innovation*, London: Idea Group, 220–34.

Kling, R., 2000, Social Informatics: A New Perspective on Social Research about Information and Communication Technologies, *Prometheus, 18*, 245–64.

Kodama, F., 1992, Technology Fusion and the New R&D, *Harvard Business Review*, 70, 70–78.

Kogut, B. and Zander, U., 1992, Knowledge of the Firm, Combinative Capabilities, and the Replication of Technology, *Organization Science*, 3, 383–97.

——, 1996, What Firms Do? Coordination, Identity, and Learning, *Organization Science*, 7, 502–18.

Koku, E. and Wellman, B., 2004, Scholarly Networks as Learning Communities: The Case of Technet, in S. Barab, R. Kling and J. H. Gray (eds), *Building Online Communities in the Service of Learning,* Cambridge: Cambridge University Press, 299–337.

Knorr-Cetina, K. D., 1999, *Epistemic Cultures: How the Sciences Make Knowledge*, Boston, MA: Harvard University Press.

Koutsopoulos, K., 1982, GIS in Greece, *Oikonomikos Tachydromos*, 21 January, 51–53 (in Greek).

Krackhardt, D. and Stern, R., N., 1988, Informal Networks and Organizational Crises, *Social Psychology Quarterly*, 51, 123–40.

Krackhardt, D. and Hanson, J., R., 1993, Informal Networks: The Company Behind the Charts, *Harvard Business Review*, 71, 104–12.

Krucken, G., 2003, Mission Impossible? Institutional Barriers to the Diffusion of the 'Third Academic Mission' at German Universities, *International Journal of Technology Management*, 25, 18–33.

Kuhn, T., 1970, *The Structure of Scientific Revolutions*, 2nd edn, Chicago: University of Chicago Press; 1st edn 1962.

Lam, A., 1997, Embedded Firms, Embedded Knowledge: Problems of Collaboration and Knowledge Transfer in Global Cooperative Ventures, *Organization Studies*, 18, 973–98.

——, 2003, Organizational Learning in Multinationals: R&D Networks of Japanese and US MNEs in the UK, *Journal of Management Studies*, 40, 673–703.

Lave, J. and Wenger, E., 1991, *Situated Learning: Legitimate Peripheral Participation*, Cambridge: Cambridge University Press.

Latour B., 1991, Technology is Society Made Durable, in J. Law (ed.), *A Sociology of Monsters: Essays in Power, Technology, and Domination*, London: Routledge, 103–31.

——, 2005, *Re-Assembling the Social*, Oxford: Oxford University Press.

Law, J., 1994, *Organising Modernity*, Oxford: Blackwell.

Layton, E. T., 1972, Mirror Image Twins: The Communities of Science and Technology in Nineteenth Century America, in G. H. Daniels (ed.), *Nineteenth Century American Science*, Evanston, IL: Northwestern University Press.

——, 1977, Conditions of Technological Development, in I. Spiegel-Rosing and Price, de Solla, D. (eds), *Science, Technology and Society: A Cross-Disciplinary Perspective*, Beverly Hills, CA: Sage, 197–222.

Lecuyer, C, 2000, Fairchild Semiconductor and its Influence, in C. M. Lee, W. F.

Miller, M. G. Hancock and H. S. Rowen (eds), *The Silicon Valley Edge*, Stanford, CA: Stanford University Press, 158–83.

Leonard-Barton, D. L., 1992, Core Capabilities and Core Rigidities: A Paradox in Managing New Product Development, *Strategic Management Journal*, 13, 111–25.

——, 1995, *Wellsprings of Knowledge: Building and Sustaining the Sources of Innovation*, Boston, MA: Harvard Business School Press.

Leonard, D. and Sensiper, S., 1998, The Role of Tacit Knowledge in Group Innovation, *California Management Review*, 40, 112–32.

Leslie, S. W., 2000, The Biggest Angel of them All: The Military and the Making of Silicon Valley, in M. Kenney (ed.), *Understanding Silicon Valley: The Anatomy of an Entrepreneurial Region*, Stanford, CA: Stanford University Press, 48–67.

Lin, N., 2002, *Social Capital: A Theory of Social Structure and Action*, Cambridge: Cambridge University Press.

Macdonald, S., 1983, High Technology Policy and the Silicon Valley Model, *Prometheus*, 1, 330–49.

——, 1995, Learning to Change: An Information Perspective on Learning in the Organization, *Organization Science*, 6, 557–68.

——, 1996, Informal Information Flow and Strategy in the International Firm, *International Journal of Technology Management*, 11, 219–32.

——, 1998, *Information for Innovation: Managing Change from an Information Perspective*, Oxford: Oxford University Press.

Macdonald, S. and Piekkari, R., 2005, Out of Control: Personal Networks in European Collaboration, *R&D Management*, 35, 441–53.

McKelvey, M., 2001, The Economic Dynamics of Software, *Economics of Innovation and New Technology*, 10, 199–236.

Mackintosh, I., 1979, Integrated Circuits: The Coming Battle, *Long Range Planning*, 12, 28–37.

Marsden, P. V., 1990, Network Data and Measurement, *Annual Review of Sociology*, 16, 435–63.

Marsden, P. V. and Campbell, K. E., 1984, Measuring Tie Strength, *Social Forces*, 63, 482–501.

Masser, I., 1992, We are Beginning to See the Creation of a European GIS Community, *Geo Info Systems*, 2, 19–27.

Masser, I. and Onsrud, H. (eds), 1993, *Diffusion and Use of Geographic Information Technologies*, Dordrecht: Kluwer Academic.

Masser, I., Campbell, H. and Craglia, M. (eds), 1996, *GIS Diffusion: The Adoption and Use of GIS in Local Government in Europe*, London: Taylor and Francis.

Merton, R. K., 1973, *The Sociology of Science: Theoretical and Empirical Investigations*, Chicago: University of Chicago Press.

——, 1988, Reference Groups, Invisible Colleges and Deviant Behaviour in Science, in H.J. O'Gorman (ed.), *Surveying Social Life*, Middletown, CT: Wesleyan University Press, 174–89.

Mitchell, J. C., 1969, *Social Networks in Urban Settings*, Manchester: Manchester University Press.

Moore, G. E., 1996, Some Personal Perspectives on Research in the Semiconductor Industry, in R. S. Rosenbloom and W. J. Spencer (eds), *Engines of Innovation*, Boston, MA: Harvard Business School Press, 165–74.

Mowery, D. C. and Nelson, R. R. (eds), 1999, *Sources of Industrial Leadership: Studies of Seven Industries*, Cambridge: Cambridge University Press.

Mytelka, L., 1995, Dancing with Wolves: Global Oligopolies and Strategic Partnerships, in J. Hagedoorn (ed.), *Technical Change and World Economy*, Cheltenham: Edward Elgar, 182–204.

Mytelka, L. and Delapierre, M., 1987, The Alliance Strategies of European Firms in the Information Technology Industry and the Role of Esprit, *Journal of Common Market Studies*, 26, 231–53.

Narula, R., 1999, Explaining the Growth of Strategic R&D Alliances by European Firms, *Journal of Common Market Studies*, 37, 711–23.

Narula, R. and Hagedoorn, J., 1999, Innovating through Strategic Alliances: Moving towards International Partnerships and Contractual Agreements, *Technovation*, 19, 283–94.

Newman, M. E. J., 2001, The Structure of Scientific Collaboration Networks, *Proceedings of National Academy of Science (USA)*, 98, 404–9.

Newman, N., 1982, Europe's Catch-up Complex, *Management Today*, October, 82–5, 134, 136, 138.

Nisbet, R., 1970, *The Social Bond: An Introduction to the Study of Society*, New York: Knopf.

Nonaka, I. and Takeuchi, H., 1995, *The Knowledge Creating Company*, Oxford: Oxford University Press.

Nonaka, I. and Konno, N., 1998, The Concept of 'Ba': Building a Foundation for Knowledge Creation, *California Management Review*, 40, 40–54.

Orlikowski, W. J. and Barley, S., R., 2001, Technology and Institutions: What Can Research on IT and Research on Organizations Learn from Each Other?, *MIS Quarterly*, 25, 145–65.

Orr, J. E., 1996, *Talking About Machines: An Ethnography of a Modern Job*, Ithaca, NY: Cornell University Press.

Osborn, R. and Hagedoorn, J., 1997, The Institutionalization and Evolutionary Dynamics of Interorganizational Alliances and Networks, *Academy of Management Journal*, 40, 261–78.

Østerlund, C., 1996, Learning Across Contexts: A Field Study of Salespeople's Learning at Work, *Skriftserie for Psykologisk Institut*, 21(1), Aarhus University, Psykologisk Institut.

Park, R. E., 1952, The Urban Community as a Spatial Pattern and a Moral Order, in R. E. Park (ed.), *Human Communities*, Glencoe, IL: Free Press, 165–77; first published 1926.

Pavitt, K., 2003, The Process of Innovation, SPRU Electronic Working Paper Series 89, available at: www.sussex.ac.uk/Units/spru/publications/imprint/sewps/sewp89/sewp89.pdf (accessed 11 April 2004).

Peterson, J., 1991, Technology Policy in Europe: Explaining the Framework Programme and Eureka in Theory and Practice, *Journal of Common Market Studies*, 29, 269–90.

Pickering, A., 1995, *The Mangle of Practice: Time, Agency and Science*, Chicago: University of Chicago Press.

Piekkari, R., Macdonald, S. and Assimakopoulos, D., 2001, In Bed with a Stranger: Finding Partners for Collaboration in the European IT Program, *Science and Public Policy*, 28, 68–78.

Pinch, T. J. and Bijker W. E., 1987, The Social Construction of Facts and Artifacts, in W. E. Bijker, T. P. Hughes and T. J. Pinch (eds), *The Social Construction of Technological Systems*, Cambridge, MA: MIT Press, 17–50.

Polydorides, N., 1993, Technology Transfer and Training Needs: The URSA-NET Experience, in I. Masser and H. J. Onsrud (eds), *Diffusion and Use of Geographic Information Technologies*, Dordrecht: Kluwer Academic, 307–16.

Popper, K. R., 1998, *The World of Parmenides: Essays on the Pre-Socratic Enlightenment*, London: Routledge.

Porter, M. E., 1980, *Competitive Strategy*, New York: Free Press.

Preece, D. and Laurila, J. (eds), 2003, *Technological Change and Organizational Action*, London: Routledge.

Price, de Solla, D., 1963, *Little Science Big Science*, New York: Columbia University Press.

Quintas, P. and Guy, K., 1995, Collaborative, Pre-competitive R&D and the Firm, *Research Policy*, 24, 325–48.

Rappa, M. W. and Debackere, K., 1992, Technological Communities and the Diffusion of Knowledge, *R&D Management*, 22, 209–20.

Ray, T., 1998, Collaborative Research in Japan and the West, in M. Hemmert and C. Oberlander (eds), *Technology and Innovation in Japan*, London: Routledge, 151–69.

Rheingold, H., 1994, *The Virtual Community*, London: Minerva.

Richardson, D. and Presley, B., 2001, MAGE 5.8, Biochemistry Department, Duke University, Durham, NC, available at: http://kinemage.biochem.duke.edu (accessed 14 Februaruy 2001).

Richardson, G. B., 1972, The Organization of Industry, *Economic Journal*, 82, 883–96.

Riordan, M. and Hoddeson, L., 1997, *Crystal Fire: The Invention of the Transistor and the Birth of the Information Age*, New York: W. W. Norton and Company.

Rogers, E. M., 1982, *Information Exchange and Technological Innovation: The Transfer and Utilization of Technical Knowledge*, Lexington, MA: Lexington Books.

——, 1987, Progress, Problems and Prospects for Network Research: Investigating Relationships in the Age of Electronic Communication Technologies, *Social Networks*, 9, 285–310.

——, 1991, The 'Critical Mass' in the Diffusion of Interactive Technologies in Organisations, in K. L. Kraemer *et al.* (eds), *The Information Systems Research Challenge*, Boston, MA: Harvard Business School Press, 245–63.

——, 1993, The Diffusion of Innovations Model, in I. Masser and H. J. Onsrud (eds), *Diffusion and Use of Geographic Information Technologies*, Dordrecht: Kluwer Academic, 9–24.

——, 2003, *Diffusion of Innovations*, 5th edn, New York: The Free Press.

Rogers, E. M. and Kincaid, D. L., 1981, *Communication Networks: Toward a New Paradigm for Research*, New York: The Free Press.

Rogers, E. M. and Larsen, J., 1984, *Silicon Valley Fever*, New York: Basic Books.

Rosen, P., 2002, *Framing Production: Technology, Culture and Change in the British Bicycle Industry*, Boston, MA: MIT Press.

Rosenkopf, L. and Tushman, M., 1998, The Coevolution of Community Networks and Technology: Lessons from the Flight Simulation Industry, *Industrial and Corporate Change*, 7, 311–46.

Rothwell, R., 1992, Successful Industrial Innovation: Critical Success Factors for the 1990s, *Research Policy*, 22, 221–39.

Rowen, H. S., 2000, Serendipity or Strategy: How Technology and Markets Came to

Favor Silicon Valley, in C. M. Lee *et al.* (eds), *The Silicon Valley Edge*, Stanford, CA: Stanford University Press, 184–99.

Sampson, S. F., 1968, A Novitiate in a Period of Change, PhD dissertation, Department of Sociology, Cornell University, Ithaca, NY.

Saxenian, A., 1990, Regional Networks and the Resurgence of Silicon Valley, *California Management Review*, 33, 89–112.

——, 1996, *Regional Advantage: Culture and Competition in Silicon Valley and Route 128*, 2nd edn, Boston, MA: Harvard University Press.

——, 2002, Transnational Communities and the Evolution of Global Production Networks: The Cases of Taiwan, China and India, *Industry and Innovation*, 9, 183–202.

——, 2003, Bay-to-bay Strategic Alliances: the Network Linkages between Taiwan and the U.S. Venture Capital Industries, *International Journal of Technology Management*, 25, 136–50.

——, 2005, From Brain Drain to Brain Circulation: Transnational Communities and Regional Upgrading in India and China, *Studies in Comparative International Development*, 40, 35–61.

——, 2006, *The New Argonauts: Regional Advantage in a Global Economy*, Boston, MA: Harvard University Press.

Saxenian, A. and Hsu, J. Y., 2001, The Silicon Valley – Hsinchu Connection: Technical Communities and Industrial Upgrading, *Industrial and Corporate Change*, 10, 893–920.

Schilling, M. A. and Phelps, C., 2004, Small World Networks and Knowledge Creation, Social Science Research Network, available at: www.ssrn.com (accessed 15 December 2005).

Schwartz, J. and McGuinness, M., 1992, *Einstein for Beginners*, Cambridge: Icon.

Scott, J., 2000, *Social Network Analysis: A Handbook*, London: Sage.

Shane, S., 2004, *Academic Entrepreneurship: University Spinoffs and Wealth Creation*, Aldershot: Edward Elgar.

Soh, P. H. and Roberts, E. B., 2003, Networks of Innovators: a Longitudinal Perspective, *Research Policy*, 32, 1569–88.

Sorensen, O. and Waguespack, D. M. (eds), 2005, Research on Social Networks and the Organization of Research and Development: An Introductory Essay, *Journal of Engineering and Technology Management*, 22, 1–7.

Spender, J., 1996, Making Knowledge the Basis of a Dynamic Theory of the Firm, *Strategic Management Journal*, 17, 45–62.

Stanley Exhibition of Cycles, 1890, *The Engineer*, 69, 107–8.

Stein, M. R., 1960, *The Eclipse of Community: An Interpretation of American Studies*, Princeton, NJ: Princeton University Press.

Steinitz, C., 1993, GIS: A Personal Historical Perspective, *GIS Europe*, 2(5–7), 19–22, 38–40, 42–5.

Swan, J., Newell, S. and Robertson, M., 1999, Central Agencies in the Diffusion and Design of Technology, *Organization Studies*, 20, 905–31.

Swan, J., Scarbrough, H. and Robertson, M., 2002, The Construction of Communities of Practice in the Management of Innovation, *Management Learning*, 33, 477–96.

Teece, D. J., 1987, Profiting from Technological Innovation, in D. J. Teece (ed.), *The Competitive Challenge*, Boston, MA: Ballinger, 185–219.

——, 1998, Capturing Value from Knowledge Assets: The New Economy, Markets for Know-how, and Intangible Assets, *California Management Review*, 40, 55–79.

——, 2001, Strategies for Managing Knowledge Assets: The Role of Firm Structure and Industrial Context, in I. Nonaka and D. J. Teece (eds), *Managing Industrial Knowledge*, London: Sage.

Thorelli, H. B., 1986, Networks: Between Markets and Hierarchies, *Strategic Management Journal*, 2, 37–51.

Tomlinson, R. F., 1988, The Impact of the Transition from Analogue to Digital Cartographic Representation, *The American Cartographer*, 15, 249–62.

Tonnies, F., 1957, *Community and Society*, ed. and trans. C. P. Loomis, East Lansing, MI: Michigan State University Press; first published 1887.

Tuomi, I., 2004, Industrial Structure and Policy Choice: Notes on the Evolution of Semiconductors and Open Source, working paper presented at an IST Knowledge Economy Workshop, available at: www.cordis.lu/ist/about/socio-eco.htm (accessed 17 June 2005).

Tushman, M., 1988, Managing Communication Network in R&D Laboratories, in M. Tushman and W. Moore (eds), *Readings in the Management of Innovation*, Cambridge, MA: Ballinger (originally in *Sloan Management Review*, Winter 1979, 37–49).

Valente, T. W., 1995, *Network Models of the Diffusion of Innovations*, Creskill, NJ: Hampton Press.

Van Maanen, J. and Barley, S. R., 1984, Occupational Communities: Culture and Control in Organizations, *Research in Organizational Behavior*, 6, 287–316.

Van Stelle, J., 2002, Centrality and Influence in the Semiconductor Industry Association, paper presented at the International Sunbelt Conference held at New Orleans, available at: www.stanford.edu/group/esrg/siliconvalley/papers.htm (accessed 17 June 2005).

Verspagen, B. and Werker, C., 2003, The Invisible College of the Economics of Innovation and Technological Change, Working Paper 21, Eindhoven Centre for Innovation Studies, Technical University of Eindhoven, Netherlands.

Von Hippel, E., 1988, *The Sources of Innovation*, Oxford: Oxford University Press.

——, 2001, Perspective: User Toolkits for Innovation, *Journal of Product Innovation Management*, 18, 247–57.

——, 2005, *Democratizing Innovation*, Boston, MA: MIT Press.

Wasserman, S. and Faust, K., 1994, *Social Network Analysis: Methods and Applications*, Cambridge: Cambridge University Press.

Watts, D. J. and Strogatz, H., 1998, Collective Dynamics of 'Small World' Networks, *Nature*, 393, 440–42.

Webber, M. M., 1964, The Urban Place and the Non-Place Urban Realm, in M. M. Webber *et al.* (eds), *Exploration into Urban Structure*, Philadelphia, PA: University of Pennsylvania Press, 79–153.

Weber, M., 1947, *The Theory of Social and Economic Organisation*, trans. A. R. Henderson and T. Parsons, London: Hodge.

Wellman, B., 1979, The Community Question: The Intimate Networks of East Yorkers, *American Journal of Sociology*, 84, 1201–31.

——, 1988, Structural Analysis: From Method and Metaphor to Theory and Substance, in B. Wellman and S. D. Berkowitz (eds), *Social Structures: A Network Approach*, Cambridge: Cambridge University Press, 19–61.

—— (ed.), 1999, *Networks in the Global Village: Life in Contemporary Communities*, Boulder, CO: Westview Press.

——, 2002, Little Boxes, Glocalization, and Networked Individualism? in T. Makoto,

P. van den Besselaar and T. Ishida (eds), *Digital Cities II: Computational and Sociological Approaches*, Berlin: Springer, 10–25.

Wellman, B. and Leighton, B., 1979, Networks, Neighborhoods and Communities, *Urban Affairs Quarterly*, 14, 363–90.

Wellman, B. and Wortley, S., 1990, Different Strokes from Different Folks: Community Ties and Social Support, *American Journal of Sociology*, 96, 558–88.

Wellman, B. and Gulia, M., 1999, Net Surfers Don't Ride Alone, in M. Smith and P. Kollock (eds), *Communities in Cyberspace*, London: Routledge, 167–95.

Wenger, E., 1998, *Communities of Practice: Learning, Meaning, and Identity*, Cambridge: Cambridge University Press.

Wenger, E., McDermott, R. and Snyder, W. M., 2002, *Cultivating Communities of Practice*, Boston, MA: Harvard Business School Press.

Winter, S. G., 1987, Knowledge and Competence as Strategic Assets, in D. Teece (ed.), *The Competitive Challenge*, Cambridge, MA: Ballinger.

Wolfe, T., 1983, The Tinkerings of Robert Noyce: How the Sun Rose on the Silicon Valley, *Esquire*, December, 346–74.

Yan, J. and Assimakopoulos, D., 2003, Knowledge Sharing and Advice Seeking in a Software Engineering Community, in L. M. Camarinha-Matos and H. A. Afsarmanesh (eds), *Processes and Foundations for Virtual Organizations*, Dordrecht: Kluwer Academic, 341–50.

——, 2006, The Small World and Scale Free Structure of an Internet Technical Community, paper presented at the 15th International Association for Management of Technology Conference, Tsinghua University, Beijing.

Author index

Subject index

Page numbers in **bold** indicate figures, graphs and tables.